Understanding Generative AI Business Applications

A Guide to Technical Principles and Real-World Applications

Irena Cronin

Apress®

Understanding Generative AI Business Applications: A Guide to Technical Principles and Real-World Applications

Irena Cronin
Savannah, GA, USA

ISBN-13 (pbk): 979-8-8688-0281-2   ISBN-13 (electronic): 979-8-8688-0282-9
https://doi.org/10.1007/979-8-8688-0282-9

Managing Director, Apress Media LLC: Welmoed Spahr
Acquisitions Editor: Celestin Suresh John
Development Editor: Laura Berendson
Coordinating Editor: Gryffin Winkler

Cover designed by eStudioCalamar

Cover image by Pete Linforth from Pixabay (www.pixabay.com)

Distributed to the book trade worldwide by Apress Media, LLC, 1 New York Plaza, New York, NY 10004, U.S.A. Phone 1-800-SPRINGER, fax (201) 348-4505, e-mail orders-ny@springer-sbm.com, or visit www.springeronline.com. Apress Media, LLC is a California LLC and the sole member (owner) is Springer Science + Business Media Finance Inc (SSBM Finance Inc). SSBM Finance Inc is a **Delaware** corporation.

For information on translations, please e-mail booktranslations@springernature.com; for reprint, paperback, or audio rights, please e-mail bookpermissions@springernature.com.

Apress titles may be purchased in bulk for academic, corporate, or promotional use. eBook versions and licenses are also available for most titles. For more information, reference our Print and eBook Bulk Sales web page at http://www.apress.com/bulk-sales.

Any source code or other supplementary material referenced by the author in this book is available to readers on GitHub (https://github.com/Apress). For more detailed information, please visit https://www.apress.com/gp/services/source-code.

If disposing of this product, please recycle the paper

*This is in memory of my husband Danny
and his incurable love of tech.*

Table of Contents

TABLE OF CONTENTS

About the Author

 **Irena Cronin** is SVP of Product for DADOS Technology which is making an app for the Apple Vision Pro that does data analytics and visualization. She is also the CEO of Infinite Retina which provides research to help companies develop and implement AI, AR, and other new technologies for their businesses. Previous to this, she worked for several years as an equity research analyst and gained extensive experience in evaluating both public and private companies.

Cronin has a joint MBA/MA from the University of Southern California and an MS with distinction in Management and Systems from New York University. She graduated with a BA from the University of Pennsylvania with a major in Economics (summa cum laude).

About the Technical Reviewer

 **Krishnendu Dasgupta** is currently the Head of Machine Learning at Mondosano GmbH, leading data science initiatives focused on clinical trial recommendations and advanced patient health profiling through disease and drug data. Prior to this role, he co-founded DOCONVID AI, a startup that leveraged applied AI and medical imaging to detect lung abnormalities and neurological disorders.

With a strong background in computer science engineering, Krishnendu has more than a decade of experience in developing solutions and platforms using applied machine learning. His professional trajectory includes key positions at prestigious organizations such as NTT DATA, PwC, and Thoucentric.

Krishnendu's primary research interests include applied AI for graph machine learning, medical imaging, and decentralized privacy-preserving machine learning in healthcare. He also had the opportunity to participate in the esteemed Entrepreneurship and Innovation Bootcamp at the Massachusetts Institute of Technology, cohort of the 2018 batch.

Beyond his professional endeavors, Krishnendu actively dedicates his time to research, collaborating with various research NGOs and universities worldwide. His focus is on applied AI and ML.

Acknowledgments

I want to thank Celestin Suresh John and the staff at Apress for the preparation of this book. I also wish to thank Carol Cox, my best friend, for helping me to focus on writing, as well as the thousands of developers currently building the next AI technologies. This book would not be here without them.

Introduction

In the burgeoning field of artificial intelligence, Generative AI stands out as a transformative force, reshaping industries and redefining the boundaries of machine creativity and functionality. *Understanding Generative AI Business Applications: A Guide to Technical Principles and Real-World Applications* serves as a comprehensive guide to the intricate world of Generative AI, exploring its technical foundations, its rapidly expanding role in business, and its profound impact on various sensory experiences.

Chapter 1 introduces readers to the realm of Generative AI, setting the stage for a journey through its capabilities and promise. **Chapter 2** delves into the core technical concepts that form the backbone of these systems, elucidating the algorithms and architectures that enable machines to generate new, original content.

In **Chapter 3**, the focus shifts to the commercial implications of Generative AI, demonstrating its potential to revolutionize business models and value chains. **Chapter 4** zooms in on text-based applications, revealing how Generative AI is powering a new generation of language models.

Chapter 5 unpacks the intricacies of transformer-based natural language processing (NLP), the architecture driving the most advanced language models today. **Chapter 6** ventures into the chatbot technology landscape, showing how conversational agents are becoming more nuanced and context-aware.

Chapter 7 presents advanced applications of text-based AI, highlighting innovative use cases and the expansion of AI's linguistic prowess. **Chapter 8** demystifies senses-based Generative AI, extending the discussion to how AI is interpreting and synthesizing sensory data.

Chapter 9 provides an in-depth look at supportive visual algorithms and computer vision, essential for understanding how AI perceives and processes visual information. **Chapter 10** discusses auditory and multisensory experiences, showcasing AI's ability to engage with the world in a holistic, human-like manner.

Chapter 11 examines autonomous AI agents, diving into the decision-making processes, data analysis, and algorithms that enable autonomy. **Chapter 12** expands on Text-Based Generative Intelligent Agents, exploring the evolution beyond traditional chatbots to more sophisticated virtual assistants.

Real-world applications and case studies come to the forefront in **Chapter 13**, illustrating the tangible impacts of Generative AI across sectors. **Chapter 14** summarizes key insights, distilling the lessons learned into actionable knowledge. Finally, **Chapter 15** reflects on the evolving landscape of Generative AI, contemplating future directions and the ethical considerations of this powerful technology.

Throughout this book, readers will gain a nuanced understanding of Generative AI, equipped with the knowledge to harness its potential and navigate its complexities. Whether you're a data scientist, data analyst, business executive, or decision-maker, this book will illuminate the path forward in the AI-generated future.

CHAPTER 1

Introduction to the World of Generative AI

Within Artificial Intelligence (AI), Generative AI stands as a beacon of innovation and creativity, marking a significant shift in how we perceive the capabilities of machines. This chapter goes into the world of Generative AI, a dynamic subset of AI that is not only redefining the boundaries of technology but also challenging our understanding of creativity and ethics in the digital age.

At the heart of Generative AI lies its foundational principle: the ability to learn from existing data and, using this knowledge, to create new, original content. This content spans a diverse spectrum, ranging from text and images to audio and complex multimedia forms. What sets Generative AI apart is its capacity to not just analyze data but to use it as a springboard for creativity, producing results that can mimic or extrapolate from the original dataset in highly inventive ways.

The journey of Generative AI is one of rapid and remarkable evolution, propelled by groundbreaking advancements in deep learning and neural network architectures. These technological strides have not only enhanced the sophistication of Generative AI models but have also expanded their potential applications. The core of these advancements lies in deep learning's ability to process and interpret vast and intricate datasets, enabling AI systems to replicate and even augment the intricate patterns and nuances found in human-generated content.

However, Generative AI is more than a technological marvel; it is an intersection of creativity and computation. It pushes the boundaries of what machines can create and achieve, ushering in a new era where AI is an active participant in creative processes. This shift from a purely analytical role to a creative collaborator has been significant, with applications ranging from art generation to composing music and authoring written content. Its role in shaping the future of human expression, innovation, and interaction

© Irena Cronin 2024
I. Cronin, *Understanding Generative AI Business Applications*, https://doi.org/10.1007/979-8-8688-0282-9_1

is undeniable and continuously unfolding. This chapter aims to provide a thorough exploration of Generative AI, covering its core concepts, historical development, and varied applications.

What Is Generative AI?

Generative AI stands at the forefront of AI, representing a dynamic and innovative subset focused on the creation of new content. This encompasses a diverse spectrum of outputs, including text, images, audio, and other multimedia forms. What distinguishes Generative AI is its foundational principle of learning from existing datasets and leveraging that knowledge to produce original, often highly creative content that closely resembles or extrapolates from the original data.

The evolution of Generative AI has been marked by significant strides, especially in the wake of groundbreaking developments in deep learning and advanced neural network architectures. These technological advancements have not only enhanced the sophistication of Generative AI models but have also broadened their potential applications. Deep learning, in particular, with its ability to process and interpret large and complex datasets, has been instrumental in enabling these AI systems to capture and replicate intricate patterns and nuances found in human-generated content.

As a field, Generative AI intertwines elements of creativity and computation, pushing the boundaries of what machines can create and achieve. It has ushered in a new era where AI is not just a tool for analysis or automation but also an active participant in creative processes. From generating art that rivals human artists to composing music, authoring written content, and creating realistic virtual environments, the capabilities of Generative AI continue to expand and evolve.

Moreover, the impact of Generative AI extends beyond artistic and creative domains. It is increasingly playing a vital role in practical and commercial applications, such as personalized content creation, generating realistic simulations for training models, and even aiding in drug discovery and material design. The versatility and potential of Generative AI make it a critical component of modern AI research and development, driving innovation and opening up new possibilities across diverse sectors.

However, the rapid advancement of Generative AI also brings with it ethical and societal challenges, particularly in areas like authenticity, intellectual property, and the potential for misuse. The ability to generate realistic content has raised questions about trust, verification, and the implications for information dissemination in an increasingly digital world. As the capabilities of Generative AI continue to grow, so does the need for careful consideration of its impact, ensuring that its development and application are guided by ethical principles and societal norms.

In essence, Generative AI represents a convergence of technology, creativity, and ethics, forming a key pillar in the ongoing journey of AI and its role in shaping the future of human expression, innovation, and interaction.

Core Concepts in Generative AI

- **Neural Networks**: At the heart of Generative AI are neural networks, which are computational models inspired by the human brain. These networks consist of layers of interconnected nodes (neurons) that process input data and produce output. The strength of these connections (weights) is adjusted during training to minimize the difference between the actual output and the desired output.

- **Deep Learning**: Deep learning is a subset of machine learning (ML) where neural networks with many layers (deep networks) are used. These networks are capable of learning complex patterns in large amounts of data, which is crucial for generative tasks.

- **Supervised vs. Unsupervised Learning**: In supervised learning, the model is trained on labeled data (input-output pairs). However, many generative models use unsupervised learning, where the model learns to identify patterns and structures in unlabeled data.

- **Generative vs. Discriminative Models**: Discriminative models learn the boundary between classes in a dataset, while generative models learn the underlying distribution of the data. Generative models are thus capable of generating new data points that are similar to the training data. (More on Generative vs. Discriminative Models later on in the chapter.)

Here is an overview of several key generative models and techniques in AI.

Generative Adversarial Networks (GANs)

- **Basic Concept**: GANs consist of two neural networks, a generator and a discriminator, which are trained simultaneously. The generator creates fake data that resembles the training data, while the discriminator tries to distinguish between real and fake data.

- **Applications**: GANs are widely used for image generation and manipulation, such as creating photorealistic images, art generation, and more recently, in deepfakes.

Variational Autoencoders (VAEs)

- **Basic Concept**: VAEs are a type of autoencoder that generates new data points. They work by encoding input data into a latent space and then decoding from this space to reconstruct the input. The variational aspect introduces a probabilistic twist, enabling the generation of new data.

- **Applications**: VAEs are used in image generation, image denoising, and as a tool for understanding high-dimensional data in a lower-dimensional representation.

AutoRegressive Models

- **Basic Concept**: These models predict future values in a sequence based on past values. Each output element is a function of previous elements.

- **Applications**: They are used in time-series prediction, text generation (like earlier RNNs and LSTMs), and speech synthesis.

Transformers

- **Basic Concept**: Transformers use self-attention mechanisms to process sequences of data, and they excel in tasks where understanding the context and relationships in data is crucial.

- **Applications**: Beyond text generation (like GPT [Generative Pretrained Transformer] series), transformers are used in translation (like Google's BERT [Bidirectional Encoder Representations from Transformers]), image recognition and generation (like Image GPT), and even in music composition.

Role of Transformers in Generative AI

Transformers, a type of model introduced in a 2017 paper titled "Attention Is All You Need" by Vaswani et al., have revolutionized the field of Generative AI, particularly in natural language processing (NLP).

- **Architecture**: The transformer architecture, pivotal in advancing neural network approaches for natural language processing and beyond, features a unique design centered around self-attention mechanisms, eliminating the need for recurrent layers. Its structure comprises an input-embedding layer that combines word embeddings with positional encodings to retain word order information. The core of the transformer includes encoder and decoder stacks, each layer within these stacks containing two sub-layers: a self-attention mechanism and a feedforward neural network, with the decoder also including an encoder–decoder attention mechanism. Residual connections and layer normalization follow each sub-layer to facilitate training stability and efficiency. The output from the decoder passes through a final linear layer and a softmax to predict the next token in a sequence. This architecture enables parallel processing of sequences, dramatically improving performance and efficiency on tasks requiring understanding of complex dependencies within data, and serves as the foundation for subsequent models like BERT and GPT, revolutionizing the machine learning landscape.

- **Self-attention Mechanism**: The key feature of transformers is the self-attention mechanism, which computes a score for how much each element in the input sequence should attend to every other element. This allows the model to capture complex relationships and dependencies in the data.

- **Applications in Generative AI**: Transformers have been used to build some of the most advanced generative models, especially for text. Models like GPT (Generative Pretrained Transformer) series by OpenAI are based on this architecture and have shown remarkable ability in generating coherent and contextually relevant text. They have also been adapted for other generative tasks such as image generation (e.g., DALL-E) and music composition.

Restricted Boltzmann Machines (RBMs)

- **Basic Concept**: RBMs are a type of stochastic neural network that can learn a probability distribution over its set of inputs.

- **Applications**: They are used for dimensionality reduction, classification, regression, collaborative filtering, feature learning, and topic modeling.

Deep Belief Networks (DBNs)

- **Basic Concept**: DBNs are composed of multiple layers of stochastic, latent variables. They are often constructed by stacking RBMs and fine-tuning the network through a process called backpropagation.

- **Applications**: DBNs are used in image and video recognition, in motion-capture data analysis, and in some cases, for generating music.

Flow-Based Generative Models

- **Basic Concept**: These models, including normalizing flows, focus on directly learning the distribution of data. They are designed to create an invertible mapping between the data and a latent space.

- **Applications**: Used for high-fidelity image generation and in scientific applications where understanding the data distribution is key.

Challenges and Future Directions

Generative AI faces several challenges:

- **Data and Computational Requirements**: Training large transformer models requires vast amounts of data and significant computational resources, making it inaccessible for many organizations and researchers.

- **Bias and Ethics**: Since these models learn from existing data, they can perpetuate and amplify biases present in the training data. Ensuring that generative models are fair and ethical is an ongoing challenge.

- **Interpretability and Control**: The interpretability and control of Generative AI models present a complex challenge, particularly as these models advance in their ability to process and generate diverse data forms like text, images, and audio. Ensuring these AI systems make decisions and produce outputs that are both safe and aligned with human values becomes increasingly difficult. The core of the challenge lies in the models' "black-box" nature, where the internal decision-making processes are not easily understood by humans. This opacity complicates efforts to predict, explain, or control the AI's behavior, raising concerns about unintended consequences or the generation of inappropriate content. As such, developing mechanisms for better understanding and guiding these AI systems is crucial for leveraging their capabilities responsibly and ethically, ensuring they contribute positively across various domains without compromising safety or ethical standards.

The future of Generative AI is likely to involve more efficient and scalable models, better mechanisms for bias mitigation, and more robust frameworks for ethical use. Innovations are likely to focus on improving efficiency, reducing biases, and enhancing the control and interpretability of these models. The integration of Generative AI with other technologies like AR and VR could lead to more immersive and personalized experiences in gaming, training simulations, and interactive media. In sectors like healthcare, generative models could aid in drug discovery and personalized medicine, while in the creative industries, they could revolutionize content creation and design.

Understanding and harnessing the potential of Generative AI, while addressing its ethical and societal implications, remains a critical area of focus for researchers, developers, and policymakers alike.

The Various Facets: Text, Senses, and Rationale

Generative AI operates across diverse modalities such as text, images, audio, video, and multimodal applications. These AIs leverage advanced models to generate new, realistic content that spans these modalities. They excel in translating between data types, enabling applications such as generating images from text descriptions or synthesizing speech from written content. Key to these systems is their ability to understand the semantic connections between varied data forms, facilitating diverse applications from content creation to enhancing augmented reality (AR) and virtual reality (VR) experiences. This technology represents a significant step forward in AI's capability to interact with and generate complex, cross-modal content, offering innovative solutions across numerous fields. A variety of AI models, including but not limited to transformer models, play critical roles in these domains. Going forward, in several chapters, there will be mention of particular Generative AI companies and what capacities they are good for. Here, there is just a summary of what types of Generative AI are good for which kind of generation and application.

Text Generation

- **Transformer Models**: GPT-4 and its predecessors are prime examples in this domain, excelling in generating human-like text for a variety of applications.

- **Other Models**: Before transformers, models like LSTM (Long Short-Term Memory) and other RNNs (Recurrent Neural Networks) were widely used for sequential data like text.

Image Generation

- **Generative Adversarial Networks (GANs)**: GANs are pivotal in image generation, capable of creating photorealistic images and artwork. They involve a generator and discriminator network that work in tandem.

- **Transformer Models**: Adaptations of transformer models like DALL-E are used for generating images from textual descriptions, demonstrating versatility in handling visual content.

Audio Generation

- **WaveNet and SampleRNN**: These models are prominent in generating realistic speech and music. WaveNet, in particular, is known for its high-quality speech synthesis.

- **Transformer Models**: Though less common in audio generation, transformers are beginning to be explored for their potential in this modality due to their efficiency in handling sequential data.

Video Generation

- **CNN-Based Models**: Convolutional Neural Networks (CNNs) have been used for video-related tasks, focusing on understanding spatial features in video frames.

- **Transformer Models**: They are being explored for their potential to handle both spatial (image) and temporal (sequence) aspects of video data, though this is a relatively new area of application.

Multimodal Applications

- **Multimodal Transformers**: Models like CLIP, DALL-E, and Gemini demonstrate the use of transformers in handling both text and images simultaneously. Additionally, in terms of audio, there are models such as AudiClip, Jukebox, and Wave2Vec 2.0.

- **Other Models**: Various customized AI models are developed to integrate and process data across multiple modalities, depending on the specific requirements of the task.

Different Generative AI models bring their unique strengths to various modalities. Transformer models, with their advanced capabilities in handling sequential and multimodal data, have made significant impacts across multiple domains. Meanwhile, models like GANs and WaveNet have been groundbreaking in their respective areas of image and audio generation. The choice of model often depends on the specific requirements of the task at hand, such as the need for understanding sequential data, generating realistic imagery, or synthesizing human-like speech. As Generative AI continues to evolve, the interplay of these diverse models is driving innovation and expanding the possibilities of what AI can create.

Historical Milestones

Major historical milestones for Generative AI were met in the 1950s–1960s, 1990s, 2000s, 2010s, and 2020s.

1950s–1960s: The Foundations of Neural Networks

The 1950s and 1960s were pivotal decades in the history of artificial intelligence and neural networks. These years laid the foundational work for what would become the field of Generative AI. Let's delve deeper into the key developments of this era.

The Perceptron (1957)

- **Invention and Impact**: The perceptron was developed by Frank Rosenblatt in 1957. It was a groundbreaking invention because it was essentially the first algorithmically described neural network. Rosenblatt's perceptron was designed to mimic the way human brains process information, laying the groundwork for modern neural networks.

- **Functionality**: The perceptron is a type of linear classifier; it makes its predictions based on a linear predictor function, combining a set of weights with the feature vector. The algorithm allowed for automatic learning of the weights given a labeled training set, which was revolutionary at the time.

- **Limitations and Criticism**: Despite its initial promise, the perceptron was later criticized, particularly by Marvin Minsky and Seymour Papert in their 1969 book, *Perceptrons*, which pointed out fundamental limitations, such as its inability to solve nonlinearly separable problems (e.g., the XOR problem). This criticism significantly slowed down neural network research for several years.

The Concept of Neural Networks (1960s)

- **Early Research**: In the 1960s, following the development of the perceptron, there was a surge in interest and research in neural networks. Scientists were intrigued by the idea of creating machines that could mimic human thought processes and learn from their environment.

- **Technological Limitations**: One of the major hurdles during this period was the lack of computational power. The complexity of neural network algorithms required more processing power and data storage capacity than was available at the time. This limitation was a significant bottleneck in advancing the field.

- **Theoretical Developments**: Despite these limitations, the 1960s saw important theoretical work. Researchers explored various network structures and learning algorithms, laying the groundwork for future advancements. This period saw the development of early multilayer networks and the exploration of nonlinear activation functions, although these ideas would not be fully realized until much later.

- **Waning Interest**: Toward the end of the 1960s, interest in neural networks began to decline. The limitations highlighted by Minsky and Papert, combined with the computational challenges and the rise of alternative approaches in AI (like symbolic AI), led to what is now known as the first "AI Winter," a period during which funding and interest in neural network research significantly decreased.

Legacy and Resurgence

- **Long-Term Impact**: The foundational work of the 1950s and 1960s set the stage for future developments in neural networks and AI. Concepts like perceptrons and early network structures would eventually evolve into more complex and capable systems with the advent of more powerful computers and improved algorithms.

- **Resurgence in the 1980s**: The field of neural networks and, by extension, Generative AI, experienced a resurgence in the 1980s with the advent of backpropagation and the increase in computational power. This resurgence marked the beginning of the modern era of AI and neural networks, leading to the sophisticated and powerful generative models we see today.

The early explorations into neural networks during the 1950s and 1960s were crucial. They represented the initial steps toward understanding and creating intelligent systems capable of learning and adapting—a vision that continues to drive AI research and development today.

1980s: Revival of Neural Networks

The 1980s marked a significant turning point in the field of artificial intelligence, particularly in the study and development of neural networks. This period is often characterized by the resurgence of interest in neural network research, primarily fueled by the introduction of the backpropagation algorithm. Let's explore this era in more detail.

Backpropagation (1986)

- **Introduction and Developers**: The backpropagation algorithm, introduced in a landmark 1986 paper by David Rumelhart, Geoffrey Hinton, and Ronald Williams, was a pivotal moment in the history of neural networks. This algorithm addressed a fundamental challenge in neural network training: efficiently adjusting the weights in multilayer networks.

- **Technical Breakthrough**: Backpropagation is a supervised learning algorithm used for training feedforward neural networks. It employs a method known as the chain rule of calculus to compute the gradient of the loss function with respect to the weights of the network. Essentially, it works by propagating the error backward through the network layers, from the output layer toward the input layers, hence the name "backpropagation."

- **Mechanism**: The process involves two main phases: the forward pass, where the inputs are passed through the network to get the output, and the error (difference between the actual and desired output) is calculated; and the backward pass, where this error is propagated back through the network, allowing the algorithm to adjust the weights in a direction that minimizes the error.

- **Impact on Neural Networks**: Backpropagation made it feasible to train deep neural networks, which have multiple hidden layers between the input and output layers. This was a significant advancement because it meant that neural networks could now learn complex patterns and perform tasks with a level of sophistication that was not previously possible.

Revival of Interest in Neural Networks

- **Overcoming Past Limitations**: The introduction of backpropagation helped to overcome some of the limitations that had been highlighted in the 1960s and 1970s, particularly regarding the training of multilayer networks. It provided a practical way to train deep neural networks, which was a major step forward from the simple perceptrons of the previous decades.

- **Renewed Research and Funding**: The success and potential of backpropagation rekindled interest in neural network research. The 1980s saw an increase in academic and industry research in this area, as well as a boost in funding. This renewed interest helped to bring neural networks back into the mainstream of AI research.

- **Broader Applications**: With the ability to efficiently train multilayer networks, researchers began to apply neural networks to a wider range of problems, from speech recognition and image processing to more complex tasks like natural language processing.

Legacy and Continuing Development

- **Foundation for Modern AI**: The advancements in neural network training in the 1980s, especially the development of backpropagation, laid the groundwork for the deep learning revolution that would follow in the subsequent decades. It set the stage for the development of more complex architectures and the emergence of convolutional neural networks (CNNs), recurrent neural networks (RNNs), and eventually, transformers.

- **Catalyst for Innovation**: The revival of neural networks in the 1980s can be seen as a catalyst that spurred a wave of innovation in AI, leading to the development of algorithms and models that are at the core of modern AI applications.

In summary, the 1980s represented a renaissance in neural network research, driven in large part by the introduction of the backpropagation algorithm. This period not only revived interest in an area of AI that had experienced a significant downturn but also set the stage for the rapid advancements in AI and machine learning that would follow in the decades to come.

1990s: Early Generative Models

The 1990s played a crucial role in the history of Generative AI, marked by significant developments in the field of neural networks and the introduction of early generative models. A key innovation during this period was the development and refinement of Boltzmann Machines and Restricted Boltzmann Machines (RBMs). These models were instrumental in advancing the understanding of how neural networks could learn and represent complex probability distributions.

Boltzmann Machines

- **Background and Development**: Boltzmann Machines, named after the physicist Ludwig Boltzmann, were developed in the mid-1980s, with their roots tracing back to the work of Geoffrey Hinton and Terry Sejnowski. They are a type of stochastic recurrent neural network.

- **Mechanism**: A Boltzmann Machine consists of a network of symmetrically connected, neuron-like units that make stochastic decisions about whether to be on or off. The connections in the network (analogous to synapses in biological brains) have weights that are adjusted during the learning process. The learning in Boltzmann Machines involves adjusting these weights to minimize the energy of the network, following principles analogous to those in statistical mechanics.

- **Energy-Based Model**: The key concept in Boltzmann Machines is the idea of an "energy" landscape, where the network learns to represent the training data by finding low-energy configurations. This is achieved through a process known as simulated annealing, which is inspired by thermodynamic systems.

Restricted Boltzmann Machines (RBMs)

- **Introduction and Evolution**: The concept of RBMs emerged as a variation of the original Boltzmann Machine. RBMs were simplified versions introduced to overcome some of the computational complexities of the full Boltzmann Machine. In an RBM, the network is divided into two layers: a visible layer (for input data) and a hidden layer (for feature detection), with no intra-layer connections, unlike the fully connected Boltzmann Machine.

- **Training and Efficiency**: RBMs are trained using a procedure called contrastive divergence, a method developed by Geoffrey Hinton. This training method was more efficient than the procedures used for traditional Boltzmann Machines, making RBMs more practical for real-world applications.

- **Generative Capabilities**: RBMs are generative models, meaning they can learn to represent and sample from the probability distribution of the input data. After training, an RBM can generate new data that resembles the training data, making it an early example of a generative model.

Impact and Applications

- **Learning Deep Representations**: RBMs were crucial in learning deep representations of data. They could be stacked to form Deep Belief Networks (DBNs), which represented a significant step forward in the ability to train deep neural networks.

- **Contributions to Deep Learning**: The developments in RBMs and their training algorithms contributed significantly to the field of deep learning. They provided insights into how deep architectures could be effectively trained and laid the groundwork for more advanced generative models.

- **Applications**: While RBMs themselves were not widely used in practical applications compared to later models, their development was key to understanding deep learning and generative models. They were explored for tasks like dimensionality reduction, feature learning, and collaborative filtering.

Legacy of Boltzmann Machines and RBMs

The development of Boltzmann Machines and RBMs in the 1990s represents a foundational period in the history of Generative AI. These models contributed to a deeper understanding of how neural networks could learn complex distributions and generate new data. The techniques and theories developed during this period paved the way for more advanced generative models and were instrumental in the resurgence of interest in deep learning in the following decades.

2000s: Advances in Deep Learning

The 2000s were a transformative decade in the field of artificial intelligence, particularly with the advances in deep learning. A pivotal development in this era was the introduction of Deep Belief Networks (DBNs) by Geoffrey Hinton and his team in 2006. This breakthrough played a crucial role in demonstrating the potential and viability of deep learning architectures, paving the way for the rapid advancement of AI technologies.

Deep Belief Networks (2006)

- **Introduction and Context**: Geoffrey Hinton, a leading figure in the field of neural networks, along with his colleagues, introduced Deep Belief Networks in 2006. This came at a time when the potential of neural networks, especially deep architectures, was not fully realized due to training difficulties.

- **Architecture of DBNs**: A DBN is essentially a stacked architecture of multiple layers of Restricted Boltzmann Machines (RBMs). Each layer in a DBN aims to learn increasingly abstract representations of the input data. The lower layers capture basic features, while the higher layers combine these features to represent more complex patterns.

- **Training Methodology**: The innovative aspect of DBNs was their training approach. The model utilized a greedy, layer-by-layer training method. Each RBM layer was trained independently in an unsupervised manner to model the distribution of its input. Once all layers were pretrained, the entire network could be fine-tuned using supervised methods like backpropagation, particularly for tasks like classification.

Impact and Contributions

- **Overcoming Deep Network Training Challenges**: Before the advent of DBNs, training deep neural networks was challenging due to problems like vanishing gradients. The layer-wise training strategy of DBNs effectively addressed this, making the training of deep networks feasible and more efficient.

- **Demonstrating the Power of Deep Learning**: DBNs were among the first models to show that deep neural networks could outperform shallower architectures, particularly in tasks involving complex data like images and speech. This was a significant proof-of-concept for the deep learning field.

- **Stimulating Further Research**: The success of DBNs reignited interest in neural networks and deep learning. It spurred a wave of research that led to the development of other deep learning models, such as deep convolutional networks and recurrent neural networks.

Broader Implications

- **Applications**: Following their introduction, DBNs were applied to various domains, including image recognition, video processing, and speech recognition. They proved particularly effective in tasks requiring the extraction of hierarchical features from raw data.

- **Influence on Subsequent Models**: The principles of layer-wise training and feature abstraction in DBNs influenced the development of other deep learning architectures. Subsequent models, like Convolutional Neural Networks (CNNs) and autoencoders, benefited from the insights gained from DBNs.

- **Revival of Neural Networks**: The introduction and success of DBNs played a significant role in the revival of interest in neural networks, marking the beginning of what is often referred to as the "deep learning revolution." This period saw rapid advancements in AI capabilities and applications.

Legacy of Deep Belief Networks

The introduction of Deep Belief Networks in the 2000s was a watershed moment in the history of artificial intelligence. DBNs not only showcased the feasibility and effectiveness of deep neural architectures but also laid a foundation for future innovations in deep learning. Their development marked the beginning of a new era in AI, characterized by an emphasis on deep, layered architectures capable of learning complex, high-level abstractions from data. The legacy of DBNs is evident in the modern landscape of AI, where deep learning is a dominant paradigm, driving advancements across numerous fields and applications.

2010s: The Rise of Modern Generative AI

The 2010s marked a period of significant advancements in the field of Generative AI, characterized by the introduction of groundbreaking models and techniques. These developments not only expanded the capabilities of AI in creating realistic, high-quality content but also paved the way for new applications and raised important ethical considerations. Let's explore some of these key developments in more detail.

Generative Adversarial Networks (GANs, 2014)

- **Introduction and Concept**: Introduced by Ian Goodfellow and his colleagues in 2014, Generative Adversarial Networks (GANs) represented a novel approach to generative modeling. A GAN consists of two neural networks: a generator and a discriminator. The generator creates data, while the discriminator evaluates it. The two networks are trained simultaneously in a competitive manner, with the generator aiming to produce increasingly realistic data and the discriminator striving to become better at distinguishing real data from fake.

- **Impact on Image Generation**: GANs quickly became famous for their ability to generate highly realistic and high-quality images. They could create photorealistic images of objects, scenes, and even human faces that were indistinguishable from real images.

- **Broader Applications**: Beyond image generation, GANs have been used in various fields such as art creation, image super-resolution, style transfer, and more. They also played a significant role in advancing the field of unsupervised learning.

Variational Autoencoders (VAEs, 2013–2014)

- **Development and Mechanics**: VAEs, developed around 2013–2014, are another important class of generative models. They are based on the framework of autoencoders but with a twist: they enforce a probabilistic distribution (usually Gaussian) on the encoded representations. This allows VAEs to not just compress data but also to generate new data similar to the input data.

- **Applications and Significance**: VAEs have been particularly influential in unsupervised learning. They are used for tasks like image generation, image denoising, and more abstract applications like learning latent representations of data which can be useful in various data analysis tasks.

Transformer Model (2017)

- **Revolution in NLP:** The introduction of the transformer model by Vaswani et al. in 2017 was a landmark event, especially for natural language processing (NLP). The transformer architecture, based on self-attention mechanisms, significantly improved the efficiency and effectiveness of processing sequential data, like text.

- **Foundation for Generative Models:** Transformers laid the groundwork for the development of powerful generative models in NLP. Models like GPT (Generative Pretrained Transformer) and BERT (Bidirectional Encoder Representations from Transformers) have their foundations in the transformer architecture and have shown remarkable capabilities in generating coherent and contextually relevant text.

GANs for Deepfakes (Mid-2010s)

- **Emergence of Deepfakes:** The mid-2010s saw the application of GANs in creating deepfakes—highly realistic and convincing digital manipulations of audio and video. This application showcased the power of GANs in creating lifelike and convincing artificial content.

- **Ethical and Societal Implications:** While the technology behind deepfakes highlighted the advancements in Generative AI, it also raised significant ethical and societal concerns. The potential for misuse of deepfake technology in misinformation, propaganda, and privacy violations brought attention to the need for responsible AI development and usage.

The 2010s were a defining decade for Generative AI, witnessing the emergence of models and techniques that pushed the boundaries of what AI could create and emulate. From the visually compelling outputs of GANs to the sophisticated language models built on transformers, this era marked a shift toward more advanced, versatile, and sometimes controversial applications of AI. These developments not only opened up new possibilities across various domains but also brought to the forefront the importance of ethical considerations and regulations in the rapidly evolving field of AI.

2020s: State-of-the-Art Developments in Generative AI

The 2020s have continued the rapid advancement of Generative AI, with several groundbreaking developments that have not only enhanced the capabilities of AI systems but have also expanded the scope of their applications. Key among these developments are the introduction of GPT-3, DALL-E, and the emergence of multimodal models.

GPT-3 and Beyond (2020)

- **Introduction of GPT-3**: OpenAI's introduction of GPT-3 (Generative Pretrained Transformer 3) in 2020 marked a significant milestone in the field of natural language processing. GPT-3 is one of the largest and most powerful language models ever created, with 175 billion parameters.

- **Capabilities**: GPT-3 demonstrated an unprecedented ability to generate human-like text. It could write creative fiction, answer questions, summarize texts, translate languages, and even generate code, among other tasks. Its performance was often indistinguishable from that of humans in various applications.

- **Impact on AI Applications**: GPT-3 significantly broadened the possibilities for AI applications. It has been used in creating advanced chatbots, enhancing content creation, automating customer service, and more. Its release also sparked discussions about the potential and limitations of large-scale language models, including concerns about bias, ethical use, and the future of human–AI interaction. We are now up to GPT-4, which was introduced in March 2023.

DALL-E (2021)

- **Innovations in Image Generation**: In 2021, OpenAI introduced DALL-E, a variant of the GPT-3 model designed to generate images from textual descriptions. DALL-E combines the concepts of transformers with image generation, enabling it to create highly detailed and contextually relevant images based on textual inputs.

- **Capabilities and Applications**: DALL-E can generate a wide range of images, from realistic photographs to artistic renditions, based on complex and sometimes abstract textual descriptions. Its ability to understand and interpret text to create coherent and often creative visual content has implications for fields such as graphic design, digital art, and advertising.

- **Advancements in Creativity and AI**: The introduction of DALL-E has opened new discussions on the role of AI in creativity and art. It raises questions about the nature of creativity and the potential for AI to become a tool for artists and designers.

Multimodal Models (2020s)

- **Rise of Multimodal AI**: The 2020s have seen the emergence of multimodal models, which are capable of understanding, interpreting, and generating content across different modalities (such as text, images, and audio). These models represent a significant advancement in AI's ability to process and integrate information from various sensory inputs, much like humans do.

- **Applications and Impact**: Multimodal models are being explored for a range of applications, including more sophisticated and interactive AI assistants, advanced content creation tools, and enhanced AI for accessibility (such as converting text to speech for the visually impaired). They are also crucial in advancing fields like autonomous vehicles and robotics, where the integration of multiple sensory inputs is essential.

- **Technical Challenges and Developments**: Developing effective multimodal models poses significant technical challenges, particularly in integrating and aligning information from different data types. Advances in this area involve improvements in neural network architectures, training techniques, and data representation methods.

The 2020s are shaping up to be a transformative decade for Generative AI, marked by significant advancements in language models, image generation, and multimodal AI. The introduction of models like GPT-3 and DALL-E has not only showcased the remarkable capabilities of AI in mimicking and augmenting human-like creativity but has also opened up new avenues for AI applications across various fields. As these technologies continue to evolve, they are likely to further blur the lines between human- and machine-generated content, raising important questions about the role of AI in society, ethics, and the future of work.

Discriminative vs. Generative Models

Generative AI is often discussed in the context of its distinction from discriminative models in the field of AI and machine learning (ML). Understanding this difference is key to appreciating the unique capabilities and applications of generative models.

Discriminative Models

Discriminative models are designed to distinguish between different types of data. They are typically used for tasks like classification and regression. These models learn the boundaries or differences between classes or categories within a dataset.

Key Characteristics

- **Learning the Decision Boundary**: Discriminative models focus on learning the decision boundary between different classes. For instance, in a binary classification problem, the model learns to differentiate between Class A and Class B.

- **Directly Predicting Outputs**: They directly predict the output or the label of an input data point. For example, given an image of an animal, a discriminative model can classify it as a "cat" or "dog."

- **Examples**: Common examples include logistic regression, support vector machines (SVMs), and most types of neural networks used in classification tasks.

Applications

- Image and speech recognition

- Text classification

- Medical diagnosis

Generative Models

Generative models, on the other hand, are designed to generate new data instances. They learn the underlying distribution of the data, enabling them to produce content that is similar to the input data they were trained on.

Key Characteristics

- **Learning Data Distribution**: Generative models learn the distribution of data points in the input space. This involves understanding the complex relationships and structures in the data.

- **Generating New Instances**: They can generate new data instances that are representative of the training data. This can include creating entirely new examples that are plausible under the learned data distribution.

- **Examples**: Notable examples include Generative Adversarial Networks (GANs), Variational Autoencoders (VAEs), certain types of Bayesian networks, and transformer models.

Applications

- Image, video, text, and music generation

- Data augmentation

- Anomaly detection (by understanding normal data distribution)

Comparison: Generative vs. Discriminative Models

- **Data Understanding**: Generative models have a deeper understanding of data distribution, while discriminative models focus on the boundaries between different categories.

- **Flexibility in Applications**: Generative models offer more flexibility in creative and generative tasks, whereas discriminative models excel in classification and prediction tasks.

- **Complexity of Training**: Training generative models is often more computationally intensive and complex, as they have to learn the entire data distribution rather than just the decision boundary.

Summary

In summary, the distinction between generative and discriminative models lies in their approach and objectives: generative models are about understanding and creating data, while discriminative models are about distinguishing and categorizing data. Both types play crucial roles in the AI landscape, with applications that suit their respective strengths and capabilities.

Generative AI, positioned at the forefront of the AI landscape, stands as a remarkable testament to the boundless potential of machines. This chapter has provided an exploration of Generative AI, shedding light on its transformative impact in this digital age.

At the heart of Generative AI lies its defining principle: the ability to assimilate knowledge from existing data and employ that knowledge to generate entirely new and original content. This content spans a diverse spectrum, encompassing text, images, audio, and even complex multimodal creations. What sets Generative AI apart is its remarkable capacity not merely to analyze data but to utilize it as a springboard for creativity, resulting in outputs that can closely mimic or ingeniously extrapolate from the original dataset.

The journey of Generative AI is one characterized by rapid and extraordinary evolution, largely driven by groundbreaking advancements in deep learning and neural network architectures. These technological leaps have not only elevated the sophistication of Generative AI models but have also broadened the horizons of their potential applications. Deep learning, in particular, has played a pivotal role by enabling AI systems to process and interpret vast and intricate datasets, thereby empowering them to replicate and augment the intricate patterns and nuances found in human-generated content.

However, Generative AI transcends mere technological marvel; it represents a convergence of creativity and computation. It pushes the boundaries of what machines can create and achieve, ushering in a new era where AI is an active participant in creative endeavors. This shift from a purely analytical role to that of a creative collaborator has yielded remarkable applications, spanning art, image and video generation, music composition, and written content authorship. Its role in shaping the future of human expression, innovation, and interaction is undeniable and continually unfolding.

In the next chapter, Generative AI's core technical concepts are reviewed, providing insights into its inner workings and performance.

CHAPTER 2

Core Technical Concepts

Generative AI represents a cutting-edge domain within the broader field of AI, distinguished by its focus on creating algorithms that can generate new data resembling existing datasets. This domain unifies principles from machine learning, statistical modeling, and computational creativity to produce algorithms that are not just analytically powerful but also creatively potent. Key to understanding Generative AI is grasping its foundational concepts, the methodologies it employs, and the mathematical foundations it rests upon. This includes a deep dive into the most prominent types of Generative AI algorithms, such as GANs, VAEs, and transformer-based models, each with unique mechanisms, strengths, and challenges.

At the heart of Generative AI lies the principle of learning data distributions. This involves sophisticated statistical and probabilistic modeling, often employing advanced techniques like Bayesian inference, density estimation, and Monte Carlo methods. These techniques enable the algorithms to capture and reproduce the complex patterns inherent in the data. Another critical aspect is dimensionality reduction and latent space modeling, where high-dimensional data is mapped onto a lower-dimensional latent space. This mapping is essential for capturing the most significant features of the data and facilitates the generation of new data instances by exploring and sampling within this latent space.

The training of generative models is intricately tied to the design of their loss functions, which govern the learning process. These loss functions strike a balance between data fidelity (ensuring generated data closely resembles real data) and other factors like diversity or smoothness in the latent space. Understanding these functions is key to comprehending how these models learn and evolve.

Regarding specific Generative AI algorithms, GANs stand out with their unique architecture comprising a generator and a discriminator, both typically structured as deep neural networks. The adversarial training process, framed as a minimax game,

© Irena Cronin 2024
I. Cronin, *Understanding Generative AI Business Applications*, https://doi.org/10.1007/979-8-8688-0282-9_2

involves the generator creating samples and the discriminator evaluating them, with each trying to outsmart the other. However, training GANs presents challenges like non-convergence and mode collapse, requiring a careful balance between the generator and discriminator.

VAEs combine autoencoder architecture with variational Bayesian methods. They function by encoding input data into a distribution in latent space and then reconstructing it, a process governed by a loss function that includes both a reconstruction loss and a regularization term, typically the Kullback–Leibler divergence. Kullback–Leibler divergence is a measure of how one probability distribution differs from another, often used in information theory and statistics to quantify the difference between two probability distributions. VAEs find applications in tasks that benefit from smooth interpolation, like style transfer, and scenarios where explicit modeling of data distribution is advantageous.

Transformer-based generative models utilize self-attention mechanisms to process sequences, making them highly effective for sequential data generation. Their ability to handle long-range dependencies makes them suitable for complex tasks in text, image, and audio generation. However, their scalability comes at the cost of significant computational resources, especially for larger models like GPT-4.

In the broader landscape of Generative AI, several challenges and future directions are evident. Training stability and efficiency remain key concerns, especially for GANs. Addressing bias and ethical considerations is crucial, given the potential for misuse in creating deepfakes and the perpetuation of biases present in training data. The quantitative evaluation of generative models is nontrivial, often requiring innovative metrics and human judgment. Furthermore, improving data efficiency and integrating these models with other machine learning techniques are ongoing areas of research.

Introduction to Algorithms

Generative AI stands at the intersection of machine learning, statistical modeling, and computational creativity. It involves sophisticated algorithms designed to generate new data instances that mimic the statistical properties of a given dataset. This advanced technical introduction delves deeper into the methodologies, mathematical foundations, and key types of Generative AI algorithms, such as GANs, VAEs, and transformer-based models, highlighting their inner workings, strengths, and limitations.

Core Principles of Generative AI

- **Learning Data Distributions**: At its core, Generative AI is about learning the probability distribution of a dataset and generating new samples from this distribution. This involves complex statistical and probabilistic modeling, often employing techniques like Bayesian inference, density estimation, and Monte Carlo methods.

- **Dimensionality Reduction and Latent Space Modeling**: Many generative models operate by mapping high-dimensional data to a lower-dimensional latent space. This mapping aims to capture the most critical features of the data, enabling the generation of new samples by exploring and sampling within this latent space.

- **Loss Functions and Optimization**: The training of generative models is governed by carefully designed loss functions, which guide the learning process. These functions often balance between data fidelity (how close the generated data is to real data) and other aspects like diversity or smoothness in the latent space.

In-Depth Look at Generative AI Algorithms

Here, GANs, VAEs, and transformer-based Generative AI models are looked at in more detail.

Generative Adversarial Networks (GANs)

- **Architecture**: Comprises a generator (G) and a discriminator (D), both typically deep neural networks. G generates samples, while D evaluates them.

- **Adversarial Training**: Involves training G to maximize the probability of D making a mistake, while D is trained to accurately distinguish between real and fake samples. This is often framed as a minimax game with a value function $V(G, D)$.

- **Challenges**: Training GANs presents several challenges, including non-convergence, mode collapse, and the delicate balance between the generator (G) and the discriminator (D). Non-convergence can hinder progress and result in unstable training. Mode collapse occurs when the generator focuses on a limited set of data patterns, reducing the diversity of generated data. Achieving the right balance between G and D is crucial; an overly powerful generator may produce unrealistic data, while a strong discriminator can impede generator improvement. To address these issues, practitioners employ various techniques like architectural modifications, alternative loss functions, regularization, and meticulous hyperparameter tuning. Hyperparameter tuning is the process of finding the optimal configuration settings for a machine learning model to achieve the best performance on a specific task, involving parameters like learning rates and batch sizes. Continuous monitoring and parameter adjustment are key to ensuring GANs generate high-quality and diverse synthetic data.

Variational Autoencoders (VAEs)

- **Framework**: Combines autoencoder architecture with variational Bayesian methods. An autoencoder architecture consists of an encoder, which compresses input data into a lower-dimensional representation, and a decoder, which aims to reconstruct the original data from this compressed representation.

- **Loss Function**: Comprises a reconstruction loss and a regularization term (often the Kullback–Leibler divergence) to ensure a well-formed latent space. The reconstruction loss measures the dissimilarity between the model's output and the input data, while the regularization term is an additional component in the loss function that controls model complexity and prevents overfitting by imposing constraints on the model's parameters. The equation of the loss function is as follows:

$L(\theta, \phi; x)$ = Reconstruction Loss + β * KL Divergence where

$L(\theta, \phi; x)$ is the total loss for a specific input x, with θ and ϕ— parameters of the encoder and decoder in the VAE.

β is a parameter that balances the importance of the reconstruction loss against the KL divergence.

Default parameter β is set to 1 in standard VAE. β adjustments can modify the characteristics of the latent space, leading to β-VAE.

- **Applications**: Suited for tasks requiring smooth interpolation, such as style transfer, and for scenarios where modeling the data distribution explicitly is beneficial. Style transfer is a technique in computer vision and image processing where the style or artistic characteristics of one image (e.g., the brush strokes of a famous painting) are applied to the content of another image, creating a new image that combines the content of one and the artistic style of another.

Transformer-Based Generative Models

- **Mechanism**: Utilizes self-attention mechanisms to process sequences, making them highly effective for sequential data generation.

- **Advantages**: Excellent at handling long-range dependencies, making them ideal for complex sequence modeling tasks in text, image, and audio generation.

- **Scalability**: While powerful, transformers require substantial computational resources, especially for large-scale models like GPT-4.

Technical Challenges and Future Directions

Training stability and efficiency are critical challenges in Generative AI, particularly for models like GANs, leading to the development of techniques such as Wasserstein loss and gradient penalty. The Wasserstein loss, also known as Earth Mover's Distance, measures the difference between real and generated data distributions in GANs, while the gradient penalty is a regularization technique in Wasserstein GANs that enforces smooth gradients in the discriminator's output to improve training stability and mitigate mode collapse. Simultaneously, the field grapples with ethical issues, ensuring models don't amplify biases and addressing the potential misuse of technology, while striving to improve data efficiency and embrace transfer learning.

Training Stability and Efficiency

Many generative models, especially GANs, face training stability issues. Techniques like Wasserstein loss and gradient penalty have been developed to address this.

Improving training efficiency and reducing computational demands are ongoing areas of research.

Bias and Ethical Considerations

Ensuring that generative models do not perpetuate or amplify biases present in training data is a significant challenge.

The potential for misuse, especially in creating realistic but fake content (deepfakes), raises ethical concerns.

Quantitative Evaluation

Quantitative evaluation of generative models is nontrivial. Metrics like Inception Score (IS) and Fréchet Inception Distance (FID) are used for images, while perplexity is used for text, but these metrics have limitations and often require supplementing with human judgment. IS and Fréchet FID are used to quantitatively assess the quality and diversity of generated images in generative models like GANs, with IS focusing on class distribution and FID considering the similarity to real data distribution in feature space. Perplexity metrics are used to quantitatively evaluate the performance of language models by measuring how well they predict sequences of words in text data, with lower perplexity values indicating better model performance.

Data Efficiency and Transfer Learning

Improving data efficiency (the ability to learn from fewer examples) and leveraging transfer learning are key areas for making these models more accessible and versatile.

In summary, Generative AI is a field that combines deep technical complexity with immense creative potential. The development and refinement of models like GANs, VAEs, and transformers are pushing the boundaries of what's possible in data generation, offering tools capable of creating everything from art to synthetic training data.

Fundamental Data Structures

The fundamental data structures in Generative AI are integral to the design and functionality of these complex systems. These data structures enable the handling, processing, and generation of diverse data forms, such as images, video, text, and audio. Understanding these structures is key to appreciating the inner workings of generative models like GANs, VAEs, and transformers.

Arrays and Tensors

- **Arrays and Matrices**: At the most basic level, Generative AI models extensively use arrays and matrices. These structures are crucial for representing data in a structured format, whether it's pixel values in images or feature representations in various layers of a neural network.

- **Tensors**: Tensors are a generalization of matrices to higher dimensions and are the cornerstone of modern deep learning frameworks like TensorFlow and PyTorch. In generative models, tensors represent not just the input and output data but also the weights and biases of the neural networks. For instance, a 4D tensor might represent a batch of images, with dimensions corresponding to batch size, height, width, and color channels.

Graphs

- **Computational Graphs**: These are used to represent the operations and computations that occur within neural networks. Each node in a graph represents an operation (like matrix multiplication or activation functions), while the edges represent tensors flowing between operations. This structure is particularly useful for efficiently computing gradients during backpropagation in training neural networks.

Queues and Buffers

- **Replay Buffers**: In some generative models, especially those incorporating reinforcement learning or temporal dynamics (like RNNs), queues and buffers are used to store and manage data. For example, experience replay buffers in reinforcement learning store previous states, actions, and rewards to sample from and learn more efficiently.

Trees

- **Search Trees**: In generative models that involve a search component, such as certain language models or decision-making models, trees are used to represent the space of possible solutions. Each node represents a state or a part of the generated sequence, and the branches represent possible next steps.

Hash Tables

- **Lookup Tables**: For efficiency, especially in models dealing with a large vocabulary (like language models), hash tables are used for quick lookup and retrieval of information, such as word embeddings or token indices.

Probability Distributions

- **Distribution Representations**: While not a data structure in the traditional sense, the representation and manipulation of probability distributions are fundamental in Generative AI. Models like VAEs explicitly encode and decode from distributions in the latent space, necessitating structures to represent and perform operations on these distributions.

Latent Space Representations

- **Encoded Feature Spaces**: Generative models often transform input data into a latent space, a lower-dimensional representation that captures essential features. The structure of this latent space, often represented as a continuous vector space, is critical for the generative capabilities of the model.

Sparse Matrices

- **Efficiency in Large Models**: For large-scale generative models, especially those handling high-dimensional data, sparse matrices are used to represent and compute operations more efficiently, reducing memory and computational requirements.

Specialized Data Structures

- **For GANs**: GANs may use specific structures to handle the dual components of the generator and discriminator, managing the flow of data between these components.

- **For Transformers**: Data structures in transformers include those for handling attention mechanisms, positional encodings, and layer outputs.

In summary, the data structures underlying Generative AI are diverse and tailored to the specific requirements of these models. They range from fundamental structures like arrays and tensors to more complex and specialized ones like computational graphs and latent space representations. These structures enable the efficient processing, learning, and generation capabilities of generative models.

An Overview of Machine Learning

To provide a more technical perspective on how ML pertains to Generative AI, it's important to review the underlying mathematical concepts, algorithms, and specific challenges and innovations in Generative AI within the broader ML context.

Machine Learning Foundations Relevant to Generative AI

Generative AI relies on statistical learning theory and probabilistic modeling, which involves estimating and sampling from complex probability distributions. Density estimation techniques, from parametric to non-parametric methods, are crucial for learning data distributions. Optimization techniques like gradient-based learning, backpropagation, and automatic differentiation are essential for training deep generative models.

Statistical Learning Theory

- **Probabilistic Modeling**: Generative AI models are deeply rooted in probability theory. They often involve estimating and sampling from complex probability distributions, typically using methods like Maximum Likelihood Estimation (MLE), Bayesian inference, or Expectation-Maximization (EM) algorithms. MLE is a statistical method that aims to find the parameters of a model that maximize the likelihood of the observed data, essentially seeking the most probable values for those parameters given the data. Bayesian inference is a statistical approach that combines prior beliefs or knowledge with observed data to update and estimate the probability distribution of uncertain parameters or hypotheses. EM algorithms are iterative optimization techniques used to estimate the parameters of probabilistic models, particularly when dealing with latent or unobserved variables, by alternately computing expected values and maximizing the likelihood of the data.

- **Density Estimation**: This is crucial in generative modeling for learning the distribution of data. Techniques vary from parametric to non-parametric methods (like kernel density estimation). Parametric density estimation is a method where a specific mathematical distribution, with a fixed set of parameters, is assumed to model the data's underlying probability distribution. Non-parametric density estimation is an approach that does not assume a specific mathematical distribution and instead estimates the data's probability distribution directly from the observed data points, often using techniques like kernel density estimation. Kernel density estimation is a technique used to create a smooth curve that represents the likelihood of different values occurring in a dataset. Imagine placing a little bump, or "kernel," on top of each data point and adding them all up to form a smooth, continuous curve that tells you how likely different values are in your data. This helps you understand the underlying pattern or distribution of your data without assuming any specific mathematical shape for it.

Optimization Techniques

- **Gradient-Based Learning**: Most generative models, especially neural network-based ones, rely on gradient-based optimization methods (e.g., Stochastic Gradient Descent (SGD), Adam optimizer). A gradient-based optimization method is an iterative technique that adjusts the parameters of a model by following the direction of the steepest increase or decrease in a mathematical function (the gradient) to find the optimal values that minimize or maximize that function. SGD is where the model's parameters are updated using a randomly selected subset of the training data in each iteration, making it computationally efficient for large datasets. The Adam optimizer is like a smart helper for training machine learning models. It adjusts the model's settings in a way that helps it learn quickly and accurately by considering the past and the current data it's working with.

- **Backpropagation and Automatic Differentiation**: These are key for training deep generative models, allowing efficient computation of gradients even in highly complex and deep networks. Backpropagation is a mathematical technique used in training neural networks that calculates and adjusts the contribution of each neuron's output error to minimize the overall error in the network by working backward from the output layer to the input layer. A neural network is a computational model inspired by the structure and function of the human brain, composed of interconnected nodes (neurons) organized in layers, used for various machine learning tasks like pattern recognition and decision-making. Automatic

differentiation is a mathematical method used to compute the rate
of change of a function with respect to its inputs, which is essential
for training complex machine learning models by adjusting their
parameters to minimize errors.

Neural Network Architectures

- **Autoencoders**: From the basis for VAEs, which are an essential
  type of generative model. Understanding the encoder–decoder
  architecture, activation functions, and regularization techniques is
  fundamental.

- **Adversarial Learning**: Central to GANs, involving a game-theoretic
  approach where two networks (generator and discriminator) are
  trained simultaneously in a competitive manner.

Generative AI Models: A Closer Look

Generative AI models, including GANs, VAEs, and autoregressive models, constitute the
heart of creative data generation, each with its intricacies and challenges. GANs involve a
dynamic interplay between generator and discriminator networks, driven by specialized
loss functions, while VAEs rely on variational inference and the Evidence Lower
Bound (ELBO) principle, a lower bound on data likelihood in variational inference.
Autoregressive models like GPT excel in sequential data generation through transformer
architecture. These models face issues of computational complexity, require specific
evaluation metrics, and raise ethical considerations, reflecting the dynamic landscape of
Generative AI.

Generative Adversarial Networks (GANs)

- **Architecture and Training Dynamics**: GANs involve a complex
  interplay between the generator and the discriminator, each typically
  a deep neural network. The generator learns to produce data by
  trying to fool the discriminator, while the discriminator learns to
  differentiate between real and fake data.

- **Loss Functions**: Understanding the different loss functions used in GANs (e.g., cross-entropy loss, Wasserstein loss) is crucial. These functions directly impact the training stability and quality of the generated data.

- **Mode Collapse and Convergence Issues**: Addressing these challenges is critical in GAN development. Techniques like mini-batch discrimination, feature matching, and Wasserstein GANs with gradient penalty have been developed to mitigate these issues.

Variational Autoencoders (VAEs)

- **Variational Inference and ELBO**: VAEs use variational inference to approximate the posterior distribution of the latent variables. ELBO is maximized to train the model, which involves a reconstruction loss and a regularization term (KL divergence).

- **Reparameterization Trick**: This technique allows for the backpropagation of gradients through stochastic nodes, crucial for training VAEs.

Autoregressive Models

- **Sequential Data Modeling**: Models like GPT (Generative Pretrained Transformer) generate data one element at a time, conditioning each new output on the previous ones.

- **Transformer Architecture**: Understanding the mechanics of self-attention mechanisms, positional encoding, and the layer structure of transformers is key to grasping how these models excel in generative tasks.

Challenges and Innovations in Generative AI

- **Sampling and Computational Complexity**: Generative models, particularly those dealing with high-dimensional data, face challenges in efficient sampling and managing computational complexity. Techniques like importance sampling, Markov Chain Monte Carlo methods, and variational approaches are employed to address these issues.

- **Evaluation Metrics**: Unlike discriminative models where accuracy or ROC–AUC (a metric that measures the ability of a classification model to distinguish between positive and negative classes, with a higher score indicating better performance) might be used, evaluating generative models often involves metrics like IS and FID mentioned previously and perceptual similarity. Understanding these metrics is crucial for assessing model performance.

- **Bias and Ethical Considerations**: Generative models can propagate or even amplify biases present in training data. Addressing these concerns requires an understanding of both the technical aspects (e.g., fair representation learning) and the ethical implications.

- **Integration with Other ML Techniques**: Generative models are increasingly being combined with other ML techniques, such as reinforcement learning for creative problem-solving or supervised learning for semi-supervised learning tasks.

In summary, the relationship between machine learning and Generative AI is intricate and multifaceted, involving advanced concepts in statistical learning, optimization, neural network architectures, and specific challenges inherent to generative modeling. Mastery of these concepts is essential for anyone delving into the field of Generative AI, as they form the backbone of the development and application of these advanced algorithms.

How Data Fuels Generative AI

The role of data in Generative AI is multifaceted and critical across various model architectures, including transformers, GANs, VAEs, and others. Each of these models interacts with data in unique ways, and understanding these interactions provides insight into the capabilities and challenges of Generative AI.

Data Representation and Preprocessing

In machine learning, data representation and preprocessing are essential steps, whether in transformers, GANs, or VAEs. Transformers tokenize and embed data, GANs preprocess images and use latent spaces, and VAEs encode input data and handle continuous data distributions.

Transformers

- **Text**: Involves tokenization and embedding.
- **Images**: Uses patching and vector embedding of image patches.
- **Positional Encodings**: Necessary for maintaining sequence order.

GANs

- **Image Preprocessing**: Involves normalization and scaling of pixel values.
- **Latent Space**: Random noise input to the generator is a critical aspect of data handling in GANs.

VAEs

- **Encoding Input Data**: Input data is encoded into a latent space representation.
- **Handling Continuous Data**: VAEs are adept at working with continuous data distributions.

Role of Data in Specific Generative Models

In various generative models like transformers, GANs, and VAEs, data is essential. Transformers use diverse data for pretraining and excel in sequence tasks, GANs use data for training dynamics and diverse output, and VAEs focus on data distribution learning and reconstruction.

Transformers

- **Large-Scale Data Training**: Requires extensive, diverse datasets for effective pretraining.

- **Sequence Generation and Attention**: Ideal for tasks requiring awareness of context and sequence.

GANs

- **Training Dynamics**: Involves a dual process where the discriminator learns from real data to guide the generator in producing realistic synthetic data.

- **Diversity and Quality of Data**: Crucial for avoiding issues like mode collapse.

VAEs

- **Learning Data Distributions**: VAEs are trained to learn the probability distribution of the input data in the latent space.

- **Data Reconstruction**: Focus on encoding and then reconstructing data, balancing fidelity and compression.

Challenges and Solutions in Data Handling

- **Computational Resources**: Transformers and large GANs demand significant computational power for training. Techniques like distributed training and model parallelism are often employed.

Data Quality and Bias

- Across all models, the quality and representativeness of data are paramount to prevent biases and ensure realistic outputs.

- Data augmentation and careful dataset curation are common strategies.

Model-Specific Data Challenges

- **Transformers**: Managing long-range dependencies and large sequence lengths.

- **GANs**: Ensuring diversity in generated data and stabilizing training.

- **VAEs**: Balancing the reconstruction quality with the effectiveness of the latent space representation.

Integrating Diverse Data Types and Sources

In modern applications, generative models are finding utility in cross-domain scenarios, like generating images from text and demanding the fusion of diverse data types. Accomplishing this integration relies on effective data fusion techniques, which enable the combination of various data modalities, including text, images, video, and audio.

Cross-Domain Applications

- Generative models are increasingly being used in cross-domain applications, such as text-to-video generation, requiring the integration of diverse data types.

Data Fusion Techniques

- Methods for combining different types of data (e.g., textual, visual, auditory) are crucial in these applications.

Future Directions in Data-Driven Generative AI

The future of data-driven Generative AI holds a focus on scalability and efficiency as models become more complex, necessitating efficient data management. Additionally, ethical considerations surrounding data sourcing and use are emerging as vital aspects of the field's future directions.

Scalability and Efficiency

- As models grow in complexity, efficient data handling and processing become even more crucial.

Ethical and Fair Use of Data

- Ensuring that the data used for training generative models is ethically sourced and used is becoming a significant consideration in the field.

Summary

In summary, data is the foundational element that powers Generative AI. The interaction between data and different generative models like transformers, GANs, and VAEs showcases a range of techniques and challenges unique to each architecture. Understanding these interactions is key to harnessing the full potential of Generative AI and addressing the technical, computational, and ethical challenges that come with it.

In conclusion, Generative AI represents a dynamic and multidisciplinary field at the intersection of machine learning, statistics, and creativity, focused on the art of data generation. Understanding its core technical concepts, including the principles of learning data distributions, dimensionality reduction, and loss function design, is essential for navigating this domain.

Prominent generative algorithms like GANs, VAEs, and transformer-based models offer diverse capabilities and applications, from image synthesis to natural language processing. However, they also present unique challenges, such as training stability, ethical considerations, and the need for innovative evaluation metrics.

As Generative AI continues to evolve, scalability, efficiency, and ethical use of data remain paramount concerns. The field's future holds promise in addressing these challenges, expanding its applications, and integrating with other machine learning techniques. In this ever-evolving landscape, a deep understanding of foundational concepts will be the compass guiding the development and application of these advanced generative algorithms. Generative AI's journey toward creativity and innovation is far from complete, promising exciting opportunities and advancements in the years to come. The business case for Generative AI is a very rich one, which the next chapter explores and elucidates.

CHAPTER 3

The Business Case for Generative AI

In recent years, AI has undergone substantial advancements, establishing itself as a pivotal element in numerous industries. This summary delves into the current state of AI within the business context, its diverse applications, the associated benefits and challenges, and a glimpse into its promising future.

AI applications in the business sector are multifaceted, revolutionizing operations, decision-making processes, and customer experiences across various domains. From enhancing customer interactions through AI-driven chatbots and virtual assistants to harnessing AI algorithms for data analytics, predictive analytics, targeted marketing, and supply chain optimization, businesses are increasingly relying on AI to streamline their operations and gain a competitive edge.

The integration of AI into business operations brings forth numerous advantages, including increased efficiency through task automation, reduced operational costs, improved decision-making through AI-driven analytics, enhanced customer satisfaction via personalized services, competitive advantages for early adopters, and the fostering of innovation by enabling the development of new products and services.

However, alongside these benefits come several challenges and concerns that businesses must address. These encompass the need for robust data protection measures to ensure data privacy, the potential for AI systems to inherit biases from training data leading to unfair outcomes, ethical considerations in decision-making processes (such as in autonomous vehicles and AI-powered healthcare), compliance with evolving AI-related regulations, and substantial investments required for implementing AI systems, including technology and staff training.

Looking toward the future, the prospects for AI in business remain promising. AI is expected to continue its expansion into various sectors with more advanced applications. Research in areas such as AI ethics and explainability is anticipated to

© Irena Cronin 2024
I. Cronin, *Understanding Generative AI Business Applications*, https://doi.org/10.1007/979-8-8688-0282-9_3

address some of the current challenges, making AI-driven decision-making more transparent and accountable. Furthermore, AI technologies like natural language processing and computer vision will become even more integrated into everyday business operations.

Transitioning to the unique realm of Generative AI, it has already had a profound impact on businesses, showcasing how it differentiates itself from general AI in a business context. Generative AI stands out for its ability to generate creative content, a capability particularly valuable in advertising, marketing, and entertainment industries. Generative AI can produce realistic images, videos, music, and text, offering businesses the opportunity to create advertisements, personalized recommendations, and even entirely new artistic works.

Personalization and customer engagement are other areas where Generative AI excels. By analyzing user data, generative models can generate tailored recommendations, product designs, or interactive experiences, increasing customer engagement and conversion rates. Additionally, Generative AI models, especially transformer-based ones like GPT, exhibit advanced language translation and localization capabilities, facilitating effective communication with diverse customer bases and international markets.

Generative AI also plays a vital role in fraud detection and security by generating synthetic data to test the robustness of security systems and identifying potential threats promptly. It revolutionizes product design and prototyping by generating a wide range of design possibilities based on parameters and constraints, resulting in more innovative and cost-effective designs.

Furthermore, Generative AI powers conversational AI and customer support, automating customer interactions through chatbots and virtual assistants, improving response times, and lowering operational costs. It enables content generation at scale, particularly valuable in content-heavy industries like media, publishing, and e-learning, where AI can produce articles, reports, and educational materials efficiently.

In research and innovation, Generative AI serves as a powerful tool, generating hypotheses, simulating experiments, and assisting scientists and researchers across various industries in exploring new ideas and solutions, thereby accelerating the pace of innovation and discovery.

Generative AI presents a range of unique capabilities that differentiate it within the business landscape. Its proficiency in content generation, personalization, content translation, security enhancement, product design, customer support automation, content scalability, research and innovation acceleration, and more makes it a versatile technology increasingly integrated into various aspects of business operations to enhance efficiency, drive innovation, and provide a competitive advantage.

Current State of AI in Business

The current state of AI in business is marked by substantial advancements and increasing integration across various industries. AI applications in business are diverse, impacting operations, decision-making, and customer experiences. Notable applications include AI-driven chatbots and virtual assistants for customer service, AI algorithms for data analytics and extracting valuable insights, predictive analytics for forecasting customer behavior, and personalized marketing campaigns.

Supply chain optimization benefits from AI's ability to predict disruptions, optimize routes, and manage inventory efficiently, while financial services utilize AI for fraud detection, algorithmic trading, and credit risk assessment. In healthcare, AI aids in medical image analysis, drug discovery, patient data management, and telemedicine, ultimately improving patient care. Manufacturing and Industry 4.0 benefit from AI-powered robots and automation systems, enhancing productivity, quality control, and maintenance.

The integration of AI in business offers several advantages, including increased efficiency through task automation, cost reduction by streamlining processes, improved decision-making through AI-driven analytics, enhanced customer experiences via personalization, and competitive advantages for early adopters. AI also fosters innovation by enabling the development of new products and services.

However, AI adoption is not without challenges and concerns. These encompass data privacy concerns when handling sensitive customer data, the potential for AI systems to inherit biases from training data leading to unfair outcomes, ethical considerations in decisions made by AI systems (e.g., autonomous vehicles and AI-powered healthcare), complexities in adhering to evolving AI-related regulations, and substantial investments required for implementing AI systems.

Looking ahead, the future of AI in business is promising, with continued expansion into various sectors and more advanced applications. Increased research in AI ethics and explainability aims to address current challenges, making AI-driven decision-making more transparent and accountable. Additionally, AI technologies like natural language processing and computer vision will become even more integrated into everyday business operations.

More detail on the current state of AI in business follows.

AI Applications in Business

AI is revolutionizing business across diverse sectors with applications such as AI-driven chatbots and virtual assistants in customer service, data analytics for informed decision-making, predictive analytics for forecasting, personalized marketing campaigns, supply chain optimization, financial services enhancements, healthcare improvements, and AI-powered automation in manufacturing. More on these are as follows:

Customer Service: AI-driven chatbots and virtual assistants are being used to enhance customer interactions. These AI systems can provide instant responses, handle routine queries, and improve customer satisfaction.

Data Analytics: AI algorithms are instrumental in processing vast amounts of data to extract valuable insights. This helps businesses make data-driven decisions, optimize processes, and identify trends.

Predictive Analytics: AI models are used to predict customer behavior, demand trends, and potential issues. This aids in inventory management, sales forecasting, and risk assessment.

Marketing and Personalization: AI is used for targeted marketing campaigns, content recommendation, and personalization. It enables businesses to tailor their offerings to individual customer preferences.

Supply Chain Optimization: AI optimizes supply chain operations by predicting disruptions, optimizing routes, and managing inventory efficiently.

Financial Services: AI algorithms are used for fraud detection, algorithmic trading, credit risk assessment, and improving customer financial experiences.

Healthcare: AI assists in medical image analysis, drug discovery, patient data management, and telemedicine, improving patient care and outcomes.

Manufacturing and Industry 4.0: AI-powered robots and automation systems enhance productivity, quality control, and maintenance in manufacturing.

Benefits of AI in Business

The integration of AI in business yields multiple advantages, including increased efficiency through task automation, cost reduction across operations, improved decision-making via AI-driven analytics, enhanced customer experiences with personalized services, a competitive edge for companies adopting AI, and the fostering of innovation by enabling new product and service development.

Here is more detail:

- **Efficiency**: AI serves as a powerful tool for streamlining operations by automating repetitive tasks, resulting in a significant reduction in manual labor, increased productivity, and a marked decrease in the likelihood of errors.

- **Cost Reduction**: By integrating AI into various business processes, companies can achieve substantial cost reductions. AI-driven automation not only reduces labor costs but also optimizes resource allocation, leading to improved cost-efficiency.

- **Improved Decision-Making**: AI empowers organizations with data-driven insights that enhance decision-making processes. By analyzing vast datasets in real time, AI provides actionable information that enables more informed and effective strategic choices.

- **Enhanced Customer Experiences**: AI's ability to offer personalized services and prompt responses leads to a noticeable improvement in customer satisfaction. This personalization fosters stronger customer relationships, higher loyalty, and increased brand reputation.

- **Competitive Advantage**: Companies that embrace AI gain a competitive edge within their respective industries. AI-powered processes and innovations enable businesses to outperform competitors, seize new opportunities, and stay at the forefront of their markets.

- **Innovation**: AI acts as a catalyst for innovation, driving the development of groundbreaking products and services. Its capacity to analyze data, identify trends, and uncover insights fosters the creation of unique solutions that can transform industries and open up new avenues for growth.

Challenges and Concerns

AI also presents challenges and concerns, such as the need for robust data protection measures to handle sensitive customer data, potential biases inherited by AI systems from training data leading to unfair outcomes, ethical dilemmas arising from AI decisions in areas like autonomous vehicles and healthcare, the complexity of adhering to evolving AI-related regulations, and the requirement for substantial investments in technology and staff training to implement AI systems.

Here is more detail:

- **Data Privacy**: Safeguarding sensitive customer data is paramount in the age of AI. Implementing robust data protection measures is essential to ensure that personal and confidential information remains secure. This includes encryption, access controls, and strict data handling protocols to mitigate the risk of data breaches. The area of Generative AI data privacy has gained much interest from investors, as it is a very important issue. A whole industry is growing around this area.

- **Bias and Fairness**: AI systems can inadvertently inherit biases present in their training data, which can result in unfair outcomes, particularly in decision-making processes. Addressing bias requires continuous monitoring, data preprocessing techniques, and the development of fairness-aware algorithms to ensure equitable results.

- **Ethical Considerations**: AI's capacity to make autonomous decisions raises ethical dilemmas, especially in fields like autonomous vehicles and AI-powered healthcare. Ensuring ethical AI involves establishing clear guidelines, accountability mechanisms, and transparency in AI systems to make responsible choices aligned with human values.

- **Regulatory Compliance**: The landscape of AI-related regulations is evolving rapidly. Businesses must navigate complex legal frameworks, privacy laws, and industry-specific regulations to ensure compliance. Staying up to date with these regulations and adapting AI practices accordingly are crucial to avoid legal issues and penalties.

- **Integration Costs**: Implementing AI systems can require substantial investments in both technology and staff training. This includes acquiring the necessary hardware and software, hiring or upskilling personnel with AI expertise, and dedicating resources to ensure successful integration and operation of AI solutions within the organization. These upfront costs can be a significant consideration for businesses.

Future Outlook

The future of AI, in general, in business is promising. AI is expected to continue its expansion into various sectors, with more advanced applications. Increased research in areas like AI ethics and explainability will address some of the current challenges. AI-driven decision-making will become more transparent and accountable. Additionally, AI technologies like natural language processing and computer vision will become even more integrated into everyday business operations.

In conclusion, AI is reshaping the business landscape, offering numerous opportunities and challenges. Its adoption is likely to increase as companies recognize the potential for improved efficiency, customer satisfaction, and competitiveness. However, ethical considerations and data privacy will remain crucial factors in the responsible deployment of AI in business.

Why Generative AI Is Different

Generative AI has had a profound impact on businesses in various distinctive ways. The unique aspects of Generative AI that set it apart from AI in general in the context of business are as follows.

Content Generation and Creativity

One of the key differentiators of Generative AI in business is its ability to generate creative content. Generative models, such as GANs, can create realistic images, videos, music, and text. This is particularly valuable in industries like advertising, marketing, and entertainment, where businesses can use AI-generated content to produce advertisements, personalized recommendations, and even entirely new artistic works.

For example, AI-generated copywriting can be used to create compelling ad campaigns, saving time and resources. However, content generation using Generative AI should be tempered with the understanding that IP and copyright issues are still not resolved by the courts.

Personalization and Customer Engagement

Generative AI plays a crucial role in enhancing customer experiences through personalization. By analyzing user data, generative models can generate tailored recommendations, product designs, or even interactive experiences. For instance, in e-commerce, Generative AI can suggest products based on individual preferences, increasing customer engagement and conversion rates. In the gaming industry, AI-driven game worlds adapt to each player's choices, providing a personalized gaming experience.

Content Translation and Localization

Generative AI models, especially transformer-based models like GPT (Generative Pretrained Transformer), have advanced language translation and localization capabilities. Businesses operating globally can use these models to translate content accurately and adapt it culturally. This ensures effective communication with a diverse customer base, facilitating market expansion and internationalization.

Fraud Detection and Security

Generative AI is also employed in business for fraud detection and security purposes. AI models can generate synthetic data to test the robustness of security systems. Additionally, they can detect anomalies and patterns in financial transactions or network traffic, helping businesses identify potential security threats and vulnerabilities promptly.

Product Design and Prototyping

In industries like manufacturing and product design, Generative AI has revolutionized the prototyping process. Design parameters and constraints can be fed into AI models to generate a wide range of design possibilities. This not only accelerates the product development cycle but also leads to more innovative and optimized designs, ultimately saving costs and improving product quality.

Conversational AI and Customer Support

Generative AI, particularly in the form of chatbots and virtual assistants, has become an essential tool for customer support. These AI systems can engage with customers in natural language, answer inquiries, and even solve problems autonomously. This automation improves response times and frees up human support agents to handle more complex issues.

Content Generation at Scale

Generative AI allows businesses to create content at scale without compromising quality. This is particularly valuable for content-heavy industries such as media, publishing, and e-learning. AI can generate articles, reports, and educational materials, significantly reducing the time and effort required to produce large volumes of content.

Research and Innovation

Generative AI is a powerful tool for research and innovation across various industries. It can generate hypotheses, simulate experiments, and assist scientists and researchers in exploring new ideas and solutions. This accelerates the pace of innovation and discovery.

In summary, Generative AI is different in the business context due to its unique ability to generate creative content, enhance personalization, improve customer engagement, and address specific business challenges such as fraud detection and security. It is a versatile technology that is increasingly being integrated into various aspects of business operations to drive efficiency, innovation, and competitive advantage.

Key Business Scenarios and Use Cases

Generative AI, with its capacity to create content and generate data, has found numerous applications across various business scenarios and use cases. Key scenarios and use cases where Generative AI makes a substantial impact on businesses are detailed here.

Content Generation and Marketing

- **Use Case**: Content creation is a time-consuming task for businesses. Generative AI can assist by producing written content, including blog posts, articles, product descriptions, and social media posts. It can also generate creative assets like images and videos for marketing campaigns.

- **Benefits**: This use case allows companies to produce high-quality, relevant content at scale, improving their online presence, search engine rankings, and customer engagement.

Personalized Recommendations

- **Use Case**: E-commerce platforms and content streaming services leverage Generative AI to provide personalized product and content recommendations to users based on their browsing and purchase history.

- **Benefits**: Personalization enhances customer satisfaction and increases sales by offering users relevant products or content, ultimately improving conversion rates.

Conversational AI and Customer Support

- **Use Case**: Businesses use Generative AI-powered chatbots and virtual assistants to provide 24/7 customer support. These AI systems can answer inquiries, assist with common issues, and guide users through processes.

- **Benefits**: Automating customer support reduces response times, lowers operational costs, and ensures consistent support quality.

Natural Language Processing and Understanding

- **Use Case**: Generative AI, especially transformer-based models, is employed for natural language processing (NLP) tasks, including sentiment analysis, language translation, and text summarization.

- **Benefits**: NLP enables businesses to gain insights from textual data, communicate with global audiences, and extract key information from large volumes of text.

Fraud Detection and Cybersecurity

- **Use Case**: Generative AI is used in cybersecurity to detect anomalies and patterns in network traffic or financial transactions. It can generate synthetic data to test the resilience of security systems.

- **Benefits**: Businesses can proactively identify security threats, protect sensitive data, and enhance the overall security posture.

Content Translation and Localization

- **Use Case**: Global businesses utilize Generative AI for accurate language translation and localization of content, including websites, apps, and marketing materials.

- **Benefits**: This ensures effective communication with international audiences, facilitates market expansion, and enhances brand perception.

Product Design and Prototyping

- **Use Case**: Generative AI assists in product design by generating design possibilities based on parameters and constraints. It is particularly valuable in industries such as manufacturing and automotive.

- **Benefits**: Accelerated product development, innovative design solutions, and cost savings are achieved through Generative AI-driven prototyping.

Data Augmentation

- **Use Case**: Generative AI can create synthetic data to augment training datasets for machine learning models. This is especially useful when labeled data is scarce.

- **Benefits**: Improved model performance, reduced bias, and increased accuracy in machine learning applications.

Gaming and Content Creation

- **Use Case**: In the gaming industry, Generative AI is used to create virtual worlds, characters, and narratives. It also generates music and sound effects.

- **Benefits**: Enhanced gaming experiences, reduced development time, and the creation of unique content that keeps players engaged.

Research and Scientific Discovery

- **Use Case**: Generative AI aids researchers by generating hypotheses, simulating experiments, and assisting in data analysis across scientific domains. An example of research LLMs and LLM-V is CheXagent by Stanford AIMI (www.marktechpost.com/2024/01/29/researchers-from-stanford-introduce-chexagent-an-instruction-tuned-foundation-model-capable-of-analyzing-and-summarizing-chest-x-rays/; https://github.com/Stanford-AIMI/CheXagent).

- **Benefits**: Accelerated research, exploration of new ideas, and potential breakthroughs in various scientific fields.

In conclusion, Generative AI has a broad range of applications across business scenarios, from content generation and personalization to cybersecurity and scientific research. Its ability to create content, provide personalized recommendations, and automate tasks makes it a valuable tool for improving efficiency, customer engagement, and decision-making in various industries.

Return on Investment (ROI) Metrics and Case Studies

ROI metrics and case studies are essential for assessing the value and effectiveness of Generative AI implementations in businesses. The following are ROI metrics commonly associated with Generative AI and real-world case studies to illustrate its impact.

ROI Metrics for Generative AI

- **Cost Savings**: Businesses often measure the ROI of Generative AI by assessing cost savings. This can include reduced labor costs due to automation and efficiency improvements. It also encompasses savings from avoiding errors and rework.

- **Revenue Increase**: Generative AI can lead to increased revenue through personalized recommendations, improved customer engagement, and more effective marketing campaigns. Tracking revenue growth directly linked to AI implementations is a crucial metric.

- **Customer Satisfaction**: Metrics such as Net Promoter Score (NPS) and customer satisfaction surveys can gauge the impact of Generative AI on customer experiences. Higher scores indicate improved customer satisfaction and loyalty.

- **Conversion Rate**: For e-commerce and content platforms, an increase in conversion rates can be a key ROI metric. Generative AI-driven personalization often results in higher conversion rates for product purchases or content consumption.

- **Time Savings**: Measuring the time saved through automation and AI-driven processes is essential. This can include reduced response times in customer support or faster content creation.

- **Quality Improvement**: Metrics related to the quality of outputs, such as content quality scores, can help quantify the ROI of Generative AI. Higher-quality content can lead to improved user engagement and conversions.

Generative AI Case Studies

Netflix: Content Recommendation

ROI Metric: Revenue Increase

    Case Study: Netflix's recommendation system, powered by Generative AI algorithms, is a prime example. The company reported that 80% of content watched on its platform is driven by recommendations. This has led to a significant increase in user engagement and subscription retention, resulting in substantial revenue growth (`www.newamerica.org/oti/reports/why-am-i-seeing-this/case-study-netflix/`).

OpenAI: ChatGPT

ROI Metric: Customer Satisfaction

    Case Study: OpenAI's ChatGPT model has been employed by businesses in customer support chatbots. One company, a major e-commerce retailer, reported a 30% increase in customer satisfaction scores after implementing ChatGPT-powered chat support. This improvement directly translated into higher customer loyalty and repeat business (`www.thoughtworks.com/en-us/insights/blog/generative-ai/can-business-trust-chatgpt`).

IBM: Watson Discovery

ROI Metric: Quality Improvement

    Case Study: IBM's Watson Discovery is used for document analysis and insights generation. A healthcare organization used Watson Discovery to analyze medical research papers, leading to more accurate and faster insights. The quality of medical recommendations improved, leading to better patient care and outcomes (`https://medium.com/mrya/ai-catalyzing-precision-medicine-for-customized-healthcare-solutions-11358590ebe8`).

McDonald's: Product Quality Initiatives

ROI Metric: Quality Improvement

Case Study: In December 2023, McDonald's announced it has plans to leverage Generative AI to expedite innovation within its equipment, identify disruptive trends affecting its business and supply chain, and simplify tasks for restaurant staff (`www.verdictfoodservice.com/news/mcdonalds-accenture-partner-ai/`).

Summary

In summary, Generative AI can deliver a substantial ROI for businesses through cost-savings, revenue increase, customer satisfaction improvement, time savings, and quality enhancement. Real-world case studies across various industries demonstrate the tangible benefits of implementing Generative AI solutions. These metrics and case studies underscore the value and potential of Generative AI in enhancing business operations and outcomes.

In conclusion, AI's impact on businesses is vast, revolutionizing operations, decision-making, and customer interactions across various sectors. It ranges from AI-powered chatbots enhancing customer service to AI algorithms driving data analytics, predictive insights, marketing, and supply chain optimization. Businesses are increasingly adopting AI to gain a competitive edge.

Integrating AI into business processes offers numerous advantages. It automates tasks, reducing manual effort and errors, leading to increased efficiency. Cost reduction follows as operational expenses are optimized. AI-driven analytics provide actionable insights, fostering innovation and enhancing customer experiences through personalization. Companies embracing AI gain a competitive advantage.

However, these benefits come with challenges. Robust data protection is crucial for handling sensitive customer data. AI systems must combat inherited biases and address ethical dilemmas, such as in autonomous vehicles and healthcare. Navigating evolving AI regulations and the need for substantial investments in technology and staff training are additional hurdles.

The future of AI in business looks promising, with expectations of expansion into various sectors. Research in AI ethics and transparency will address challenges, making AI-driven decision-making more accountable. AI technologies like natural language processing and computer vision will continue to integrate into daily operations.

Generative AI, a subset with unique capabilities, has already left its mark on businesses. It excels in content creation, benefiting advertising, marketing, and entertainment. Personalization and customer engagement are strong suits, while advanced language translation and security enhancement further enhance its value.

Generative AI also powers conversational AI, content scalability, and innovation acceleration. It plays a pivotal role in product design, automating customer support, and generating content at scale.

AI's recent progress has made it a cornerstone of change in business. While presenting challenges, it promises to continue shaping the business world. Generative AI, with its distinct capabilities, further augments business operations, making it an indispensable tool for success in the digital age. The next chapter goes into more detail on one aspect of Generative AI where it is showing great impact—text-based Generative AI.

CHAPTER 4

The World of Text-Based Generative AI

Text-based Generative AI, nestled within the vast landscape of AI, represents a remarkable convergence of technology and linguistics. This intriguing domain is dedicated to the art and science of crafting human-like textual content by discerning intricate patterns and assimilating knowledge from extensive datasets. This technological marvel is more than just a mere tool—it is a catalyst for innovation that holds the potential to reshape communication, creativity, and comprehension as we know it.

At its core, text-based Generative AI harnesses the power of language in unprecedented ways. It stands as a testament to the incredible strides made in the field of natural language processing (NLP), where algorithms are meticulously trained to understand, interpret, and generate human language with a level of sophistication and nuance that was once unimaginable. This technology has transcended the boundaries of mere automation; it has unlocked the ability to generate text that not only adheres to grammatical rules but also resonates with the subtleties of context, tone, and creativity.

Categories and Subtypes

NLP, as a fundamental branch of text-based Generative AI, enables machines to comprehend, interpret, and generate human language. Subcategories of NLP include text generation and sentiment analysis, which find applications in chatbots, content creation, and sentiment assessment. Machine translation is another vital category, featuring subtypes like Neural Machine Translation (NMT) and language generation, while text summarization aids in distilling crucial information from lengthy documents through extractive and abstractive methods.

© Irena Cronin 2024
I. Cronin, *Understanding Generative AI Business Applications*, https://doi.org/10.1007/979-8-8688-0282-9_4

NLP

NLP is a fundamental category of text-based Generative AI that enables machines to understand, interpret, and generate human language. It serves as a vital bridge between human communication and computational capabilities, opening up a world of possibilities for improving communication, automating tasks, and extracting insights from text data. More focus on NLP can be found in Chapter 5, "Unpacking Transformer-Based NLP." Subtypes of NLP include the following:

Text Generation: Text generation within NLP focuses on machines producing coherent, grammatically correct, and contextually relevant text. This is achieved through various techniques, including deep learning and language models. Practical applications include the development of chatbots, automated content generation for blogs and websites, and even the creation of compelling narratives or stories. For businesses, text generation streamlines content creation processes, saving time and resources while ensuring consistency and quality.

Sentiment Analysis: Sentiment analysis is a specialized area of NLP that allows machines to assess the sentiment or emotional tone conveyed by text, whether it's positive, negative, or neutral. It's invaluable for businesses seeking insights from customer feedback, product reviews, or social media interactions. Automated sentiment analysis helps companies understand customer sentiment, track brand perception, and make informed decisions to improve products and services. It also plays a role in automating customer support by identifying customer frustrations or dissatisfaction.

Machine Translation

Machine translation is a critical category within text-based Generative AI that addresses the challenge of translating text from one language to another. It has broad applications in a globalized world where communication across language barriers is crucial.

Subtypes include the following:

Neural Machine Translation (NMT): NMT represents a significant advancement in machine translation, leveraging deep learning models to improve the accuracy and fluency of translations. It's extensively used by businesses operating on a global scale for translating content, documents, and communication, enabling effective cross-border communication and market expansion.

Language Generation: Language generation goes beyond mere translation; it focuses on generating text in a target language that is not only accurate but also contextually appropriate and culturally sensitive. This is particularly important for businesses looking to engage with diverse international audiences, ensuring that translated content resonates with local customs, idioms, and preferences, facilitating effective communication and cultural sensitivity.

Text Summarization

Text summarization is a critical aspect of NLP that enables the extraction of essential information from lengthy documents or articles, thereby improving information retrieval and comprehension. Subtypes include the following:

Extractive Summarization: Extractive summarization involves selecting the most relevant sentences or phrases from a text while preserving the original wording. This technique aims to provide a concise summary that retains the essence of the original content. It is widely used in scenarios where maintaining the fidelity of the original content is essential, such as news articles and legal documents.

Abstractive Summarization: Abstractive summarization takes a more creative approach by interpreting the source text and generating new sentences that convey the main ideas concisely. This method allows for greater flexibility in summarization, as it can capture complex ideas and present them in a condensed form, making it useful in contexts where brevity and clarity are paramount.

NLP is a rich and evolving field, encompassing various subcategories with wide-ranging applications. From enhancing communication through text generation to gaining insights from sentiment analysis, facilitating cross-cultural interactions with machine translation, and improving information retrieval with text summarization, NLP continues to shape the way we interact with and leverage human language in the digital age.

The Business Value Proposition

Generative AI, particularly in the realm of text-based applications, is not just a technological advancement; it's a game-changer for businesses across various industries. This innovative technology has opened up a world of possibilities, from automating content generation to revolutionizing customer support and personalization.

Efficiency and Innovation Unleashed

One of the most significant contributions of Generative AI to businesses is its capacity to streamline operations and foster innovation. Content generation, a time-consuming task, can now be automated, freeing up resources for more creative endeavors. Customer support has taken a giant leap forward with AI-powered chatbots offering round-the-clock assistance. This not only enhances customer satisfaction but also allows human teams to focus on complex tasks. Personalization, driven by AI insights, has become a key differentiator, providing tailored experiences that drive engagement and loyalty. These innovations are not limited to one sector; they span industries, enabling organizations to adapt and thrive in the digital age.

Breaking Down Barriers and Expanding Horizons

Generative AI has emerged as a bridge between languages and cultures, facilitating global expansion. Machine translation, a subset of Generative AI, has shattered language barriers, enabling businesses to reach international audiences with ease. It's not just about translation; it's about conveying context and cultural nuances, fostering meaningful communication. This newfound global reach opens up doors to untapped markets and international collaborations. Moreover, Generative AI's ability to analyze vast amounts of textual data unlocks insights, helping businesses make data-driven decisions, identify market trends, and respond to customer feedback effectively. Whether it's in finance, healthcare, or social media analytics, Generative AI is a valuable ally in the quest for actionable insights.

Adaptation Across Industries

Generative AI's adaptability is its superpower. From healthcare to gaming and crisis communication to financial analysis, it has found applications in diverse sectors. It assists in educational tools, legal research, and even environmental monitoring. The technology's ability to generate code, transcribe speech, and improve quality assurance in software development showcases its versatility. In essence, Generative AI isn't just a tool; it's a transformative force that empowers businesses to innovate, connect with their audience on a deeper level, and thrive in a rapidly evolving landscape.

Generative AI isn't just a buzzword; it's the future of business. Its impact spans industries, reshaping how we create, communicate, and operate. Businesses that harness its potential stand to gain a competitive edge, delivering personalized experiences, breaking down language barriers, and making data-driven decisions. As Generative AI continues to evolve, its transformative power will only grow, cementing its place as a cornerstone of modern business strategies.

A deeper dive into a number of text-based Generative AI applications can be found in the next chapter, Chapter 5, "Unpacking Transformer-Based NLP." Some of the key uses of text-based Generative AI include the following:

Content Ideation: Content creators and marketers can use AI to brainstorm ideas for creative content, ad campaigns, slogans, and brand names.

Content Generation: Text-based Generative AI is employed to automate the creation of written content. It can generate articles, blog posts, product descriptions, and marketing materials efficiently and at scale. This is particularly useful for content marketing and publishing industries.

Customer Support: Chatbots and virtual assistants powered by NLP are used for instant customer support. They can answer questions, provide information, and resolve issues 24/7, improving customer service and reducing response times.

Personalization: Text-based Generative AI analyzes user data to personalize content and recommendations. It's used extensively in e-commerce, streaming services, and online advertising to tailor experiences for individual users, enhancing engagement and satisfaction.

Language Localization: Machine translation, a subset of text-based Generative AI, is essential for breaking down language barriers and facilitating global communication. It's used by businesses expanding internationally to translate content and communication materials.

Language Generation: This aspect of text-based AI goes beyond translation and aims to generate text in a target language that is contextually appropriate and culturally sensitive. It's used in creative content adaptation, such as literature and advertising, where maintaining cultural nuances is crucial.

Real-time Language Translation: AI-driven translation services are used for real-time language translation during conferences, meetings, and international events, facilitating communication among multilingual attendees.

Data Analysis: Text-based Generative AI can analyze large volumes of textual data for insights, sentiment analysis, and market trends. This is valuable for businesses in various industries, including market research, finance, and social media analytics.

Efficient Documentation: In industries where documentation is critical, such as healthcare and legal sectors, text-based Generative AI assists in generating reports, manuals, and documentation with high precision and efficiency, reducing human errors.

Text Summarization: Text summarization, a component of text-based Generative AI, condenses lengthy documents, articles, or reports into concise summaries, aiding in knowledge management and faster information retrieval.

News Reporting: Some news agencies use text-based Generative AI to automate the generation of news articles or to assist journalists in fact-checking and information gathering.

Legal Research: Legal professionals use text-based Generative AI for legal research, document review, and contract analysis, saving significant time and reducing the risk of missing critical information.

Legal Drafting: Lawyers and legal professionals use AI to draft legal documents such as contracts, wills, and patent applications, ensuring accuracy and compliance with legal standards.

Compliance and Regulatory Reporting: Businesses in highly regulated industries, such as finance and healthcare, use AI to ensure compliance with industry regulations and generate detailed compliance reports.

Teaching Tools: Text-based Generative AI is used in educational applications, including language learning and automated essay grading, where it can provide personalized feedback and assessments.

Academic Assistance: AI-powered educational tools assist students and researchers by providing explanations, generating citations, and suggesting relevant sources for academic papers and projects.

Healthcare: In healthcare, it can assist in analyzing medical records and research articles, aiding in diagnosis, treatment, and research. This includes clinical report generation from clinical biomarkers, X-ray diagnostics, and the creation of formatted clinical notes.

Mental Health Support: AI chatbots are used for mental health support, offering a listening ear and providing resources and coping strategies to individuals in need.

Pharmaceutical Research: In the pharmaceutical industry, AI helps in analyzing medical literature and research data to identify potential drug candidates and research areas.

Government and Diplomacy: Text-based Generative AI can help government agencies in analyzing public sentiment and diplomatic communication, facilitating better decision-making and diplomacy.

Creative Writing Assistance: Authors and content creators sometimes use AI tools for brainstorming ideas, generating content outlines, and overcoming writer's block.

Market Research: Text-based AI helps businesses analyze customer feedback, product reviews, and social media comments to gain insights into market trends and customer sentiments.

Social Media Management: Text-based AI can assist social media managers by suggesting content ideas, writing engaging posts, and even responding to comments and messages in a personalized manner.

Social Listening and Brand Monitoring: Brands use text-based Generative AI to monitor social media platforms and online forums, tracking mentions and sentiments related to their products or services to make informed marketing and product development decisions.

Email Automation: Businesses use AI-powered tools to automate email responses, send personalized email campaigns, and categorize and prioritize incoming emails based on content analysis.

Academic Assistance: AI-powered educational tools assist students and researchers by providing explanations, generating citations, and suggesting relevant sources for academic papers and projects.

Financial Analysis: In the finance sector, text-based AI is employed to analyze news articles, financial reports, and social media chatter to identify trends, sentiment shifts, and potential investment opportunities.

HR and Recruitment: HR departments use Generative AI for automated resume screening, job candidate matching, and even drafting job descriptions and offer letters.

Virtual Assistants: AI-driven virtual assistants like Siri and Alexa utilize text-based Generative AI to provide responses and perform tasks based on user voice commands.

Gaming: AI-powered chatbots and virtual characters enhance gaming experiences by providing natural language interactions and assisting players with in-game tasks and information.

Crisis Communication: During crises or emergencies, AI chatbots can provide real-time updates, answer frequently asked questions, and offer guidance to the public, improving communication and reducing panic.

Code Generation: AI-powered code generators assist software developers in automatically generating code snippets or even entire programs based on high-level descriptions or requirements.

Speech-to-Text Transcription: Generative AI-driven speech recognition technology converts spoken language into written text, enabling transcription services, automated closed captioning, and voice-controlled applications.

Quality Assurance and Testing: AI can automatically generate test cases and perform quality assurance testing on software applications, identifying bugs and vulnerabilities.

Environmental Monitoring: AI analyzes environmental data from sensors and satellites, helping in monitoring climate change, predicting natural disasters, and managing natural resources.

Food and Recipe Creation: AI can suggest recipes based on available ingredients, dietary preferences, and nutritional requirements, making meal planning more convenient and personalized.

Tourism and Travel Planning: Generative AI assists travelers by recommending destinations, accommodations, and itineraries based on their preferences and budget.

Summary

These myriad applications reflect the adaptability and growing significance of text-based Generative AI across various sectors, demonstrating its potential to drive efficiency, innovation, and improved user experiences in a wide array of contexts.

It's clear that text-based Generative AI has emerged as a transformative force for businesses. Its ability to streamline operations and foster innovation is reshaping the way organizations operate in the digital age. With the automation of content generation and the deployment of AI-powered customer support, businesses are achieving greater efficiency and enhancing customer satisfaction. Personalization, driven by AI insights, has become a key differentiator, delivering tailored experiences that drive engagement and loyalty, spanning industries and enabling organizations to adapt and thrive.

Furthermore, Generative AI breaks down language barriers and facilitates global expansion. Machine translation conveys context and cultural nuances, fostering meaningful communication and opening doors to untapped markets and international collaborations. The technology's data analysis capabilities unlock insights, aiding in data-driven decision-making, trend identification, and effective response to customer feedback. Its adaptability spans various sectors, assisting in education, legal research, environmental monitoring, healthcare, mental health support, pharmaceutical research, government analysis, creative content generation, and much more.

In conclusion, Generative AI's transformative impact on industries across the board demonstrates its potential to drive efficiency, innovation, and improved user experiences in a wide array of contexts. As technology continues to evolve, its role as a cornerstone of modern business strategies will only grow, providing a competitive edge and enabling businesses to thrive in an ever-evolving landscape.

In the next chapter, transformer-based NLP is reviewed in much greater detail, including specific business applications.

CHAPTER 5

Unpacking Transformer-Based NLP

The evolution of NLP has been significantly marked by the advent of transformer-based models. These revolutionary models, which include the GPT variants, have ushered in a new era of language understanding and generation, redefining the capabilities of AI in dealing with human language. Prior to the rise of transformers, NLP predominantly relied on various other architectures, such as Recurrent Neural Networks (RNNs) and Convolutional Neural Networks (CNNs), each tailored for specific aspects of language tasks. However, transformers brought a game-changing concept to the table with their dynamic attention mechanisms and contextual word embeddings, allowing them to effectively capture contextual information and dependencies, regardless of the distance between words in a sentence.

At the heart of transformer-based NLP models are two essential components: the self-attention mechanism and the stacked transformer blocks. The self-attention mechanism enables the model to weigh the significance of each word or token in a sequence concerning all others, giving it the ability to capture nuanced relationships within text data and eliminating the limitations of fixed context windows seen in earlier NLP architectures. Stacked transformer blocks, equipped with multi-head self-attention and feedforward neural networks, create a deep architecture capable of learning intricate patterns and representations from text data. These transformer-based models have found applications in diverse domains, including customer service chatbots providing round-the-clock assistance, sentiment analysis for social media analytics, automated content generation for various industries, drug interaction prediction in pharmaceutical research, and legal document analysis for efficient summarization and analysis.

As transformer-based NLP models continue to advance and adapt, they are reshaping the landscape of AI and language understanding. Their remarkable ability

© Irena Cronin 2024
I. Cronin, *Understanding Generative AI Business Applications*, https://doi.org/10.1007/979-8-8688-0282-9_5

to capture contextual nuances, coupled with their scalability and versatility, positions them at the forefront of NLP research and applications. In a world where language is the bridge to human–computer interaction, these models are pioneering a future where machines understand and respond to human language with unprecedented accuracy and sophistication, driving innovation across industries and unlocking new possibilities in communication and automation.

Anatomy of Transformer Models

Not all NLP models are transformer-based. While transformer models have gained significant popularity in the NLP field due to their effectiveness in capturing contextual information, there are various NLP models and approaches that predate transformers and are still relevant today. The following are some examples that have been previously detailed:

Recurrent Neural Networks (RNNs): Before transformers, RNNs were widely used in NLP for tasks like text classification, language modeling, and sequence-to-sequence tasks. They have a sequential nature, making them suitable for tasks involving sequential data.

CNNs: CNNs, primarily known for image processing, have also been adapted for text-based tasks, particularly for tasks like text classification and sentiment analysis. They can capture local patterns in text data.

Hidden Markov Models (HMMs): HMMs were used for various NLP tasks, especially in speech recognition and part-of-speech tagging.

Word Embedding Models: Models like Word2Vec and GloVe are not transformer-based but are crucial for representing words as dense vectors. These word embeddings are still widely used in NLP tasks and are often used in conjunction with transformer models.

Statistical Models: Traditional statistical models like Naive Bayes and Conditional Random Fields (CRFs) are used in NLP for tasks like text classification and named entity recognition.

Rule-Based Approaches: Rule-based systems are used for specific NLP tasks where explicit rules can be defined, such as information extraction or grammar parsing.

While transformers, particularly models like BERT, GPT, and their variants, have achieved state-of-the-art performance in many NLP tasks and have become dominant in recent years, the NLP field is diverse, and different models and approaches continue to play important roles depending on the specific task and requirements. Researchers and practitioners choose the model that best suits their needs and the characteristics of the data they are working with.

Role in Natural Language Understanding

Transformer-based NLP models have become the linchpin of Natural Language Understanding (NLU). Their remarkable ability to capture contextual information and relationships between words and tokens has led to significant breakthroughs in various NLU tasks.

Focusing on transformer-based NLP is justified because transformer models have brought about a significant breakthrough in NLP for several compelling reasons:

Attention Mechanism: Transformers introduced the self-attention mechanism, which allows the model to weigh the importance of each word or token in a sequence concerning all other words in that sequence. This dynamic attention mechanism enables the model to capture long-range dependencies and contextual information effectively, eliminating the limitations of fixed-size context windows present in earlier NLP architectures.

Contextual Word Embeddings: Unlike traditional word embeddings like Word2Vec or GloVe, which represent words as fixed vectors regardless of context, transformers generate contextual word embeddings. This means that the meaning of a word can change based on its surrounding context, allowing the model to capture nuances, polysemy, and word sense disambiguation effectively.

Pretraining and Fine-tuning: Transformers are often pretrained on massive corpora of text data in an unsupervised manner. During pretraining, they learn general language understanding, grammar, and a broad spectrum of world knowledge. This pretrained model can then be fine-tuned on specific NLP tasks with comparatively small task-specific datasets, resulting in impressive performance across a wide range of tasks.

State-of-the-Art Performance: Transformer-based models, particularly variants like BERT, GPT-3, and their successors, have consistently achieved state-of-the-art performance on numerous NLP benchmarks and NLU tasks. They have outperformed previous models in tasks such as text classification, language generation, machine translation, question answering, and more.

Multilingual NLP: Transformers have also excelled in multilingual NLP. Models like mBERT and XLM-R demonstrate the capacity to understand and generate text in multiple languages, breaking language barriers and enabling cross-lingual applications.

Transfer Learning: Transformer-based models are highly adaptable and can be fine-tuned for various NLP tasks with minimal labeled data, making them versatile and cost-effective for businesses.

Continual Innovation: The field of transformer-based NLP continues to evolve rapidly, with ongoing research leading to increasingly powerful models and techniques. This continual innovation makes it an exciting and fruitful area for further exploration and development.

At its core, a transformer model consists of two fundamental components:

Self-Attention Mechanism

The self-attention mechanism is the backbone of the transformer architecture. It enables the model to weigh the importance of each word or token in a sequence concerning all other words in that sequence. This dynamic attention mechanism allows the model to focus on relevant words while processing a given word, capturing contextual information effectively. It eliminates the limitations of fixed-size context windows present in earlier architectures.

Transformer Blocks (or Layers)

A transformer model is composed of multiple stacked transformer blocks or layers. Each layer consists of a multi-head self-attention mechanism and feedforward neural networks. The multi-head attention mechanism allows the model to attend to different parts of the input sequence simultaneously, enhancing its ability to capture diverse relationships and dependencies within the text.

Additionally, residual connections and layer normalization are employed to stabilize training and facilitate the flow of gradients during optimization.

These transformer blocks are stacked atop each other, creating a deep architecture that can learn intricate patterns and representations from text data. The output of the final layer is used for various NLP tasks, such as text classification, language modeling, translation, and more.

Overall, transformer-based NLP models have significantly advanced the state of the art in NLP, offering a powerful toolset for a wide range of applications, from text understanding and generation to translation and sentiment analysis. Their versatility and effectiveness have made them a dominant force in the NLP landscape and have revolutionized the field of Natural Language Understanding. Their intricate architecture, featuring self-attention mechanisms and stacked transformer blocks, allows them to capture contextual information effectively. These models play a pivotal role in pretraining and fine-tuning for a wide range of NLP tasks and have propelled the state of the art in NLU, making them a cornerstone of modern NLP research and applications.

Business Applications: Customer Service, Analytics, and More

Transformer-based NLP has emerged as a transformative technology with versatile applications. In customer service, it powers virtual assistants and chatbots that provide round-the-clock support, ensuring quick and consistent responses. In social media analytics, transformer-based NLP performs real-time sentiment analysis at scale, offering granular insights into customer opinions. It also enables automated article writing, enhancing content creation efficiency and consistency. In pharmaceuticals, the technology accelerates drug interaction prediction, improving safety and cost-effectiveness. Legal professionals benefit from efficient document analysis and summarization, aiding decision-making and risk assessment. Overall, transformer-based NLP has become an indispensable tool across industries, driving innovation and efficiency in diverse applications.

Customer Service

Here is an example of how transformer-based NLP is used in customer service (more detail on chatbots can be found in the next chapter, Chapter 6, "Exploring Chatbot Technologies"):

Virtual Assistants and Chatbots: Many businesses employ virtual assistants and chatbots powered by transformer-based NLP models to enhance their customer service operations. These virtual assistants can be integrated into websites, mobile apps, or messaging platforms to provide instant and automated support to customers.

Use Case: Let's say you visit an e-commerce website and have a question about a product. Instead of waiting for human customer support, you can start a chat with the virtual assistant. You type in your query, such as "What are the specifications of the latest smartphone?" The virtual assistant, built on a transformer-based NLP model, understands your query, retrieves relevant information from the product database, and provides you with detailed specifications, pricing, and availability information—all in a conversational and user-friendly manner.

Benefits:

- **24/7 Availability**: Virtual assistants powered by transformer-based NLP are available round-the-clock, ensuring that customers can get assistance at any time, even outside of regular business hours.

- **Quick Responses**: They can provide quick and accurate responses to frequently asked questions, reducing response times and improving user experience.

- **Scalability**: These systems can handle a large volume of customer inquiries simultaneously, allowing businesses to scale their customer support without hiring additional human agents.

- **Consistency**: Virtual assistants maintain a consistent level of service quality and information accuracy, reducing the risk of human errors.

Transformer-based NLP models, with their natural language understanding capabilities, enable these virtual assistants to engage in meaningful and context-aware conversations with customers, addressing their queries, resolving issues, and providing valuable information effectively. This application of transformer-based NLP not only enhances customer service but also contributes to cost savings and improved customer satisfaction.

Sentiment Analysis for Social Media Analytics

Use Case: Imagine you work for a social media analytics company, and one of your clients is a retail brand interested in understanding customer sentiment about their products on social media platforms like Twitter. They want to analyze the sentiment of thousands of tweets mentioning their brand and products to gain insights into customer opinions.

How Transformer-Based NLP Is Applied

Data Collection: You collect a large dataset of tweets mentioning the brand and products, spanning several months.

Data Preprocessing: The text data undergoes preprocessing, which includes tokenization, removing stopwords, and handling special characters and emojis.

Sentiment Analysis Model: You use a transformer-based NLP model, such as GPT-4, pretrained on a vast amount of text data, to perform sentiment analysis. This model has learned to understand context and nuances in language.

Sentiment Classification: The transformer model is used to classify each tweet into sentiment categories like positive, negative, or neutral. It understands the context and tone of the text, allowing for accurate sentiment classification.

Visualization and Reporting: The results are visualized in dashboards and reports. You can provide your client with insights into trends, sentiment shifts, and specific product-related sentiment. For instance, you might find that there is a spike in negative sentiment around a particular product feature.

Benefits:

- **Granular Insights**: Transformer-based NLP models can provide fine-grained sentiment analysis, allowing businesses to understand not only overall sentiment but also specific aspects that drive customer opinions.

- **Real-Time Analysis**: Social media analytics using transformer-based NLP can be performed in real time, enabling brands to respond promptly to emerging trends or address customer concerns.

- **Scalability**: These models can analyze vast volumes of social media data efficiently, making them suitable for large-scale analytics projects.

- **Customization**: You can fine-tune the model to understand industry-specific language and domain-specific sentiment expressions.

Transformer-based NLP models excel in sentiment analysis and can be a valuable tool for businesses to gauge customer sentiment, track brand reputation, and make data-driven decisions to improve products and services.

Automated Article Writing

Use Case: Imagine you run a content marketing agency, and you have a client in the travel industry who needs a constant stream of travel-related blog posts to engage their audience. To meet their content demands efficiently, you use transformer-based NLP for automated article writing.

How Transformer-Based NLP Is Applied

Topic Selection: You start by selecting a travel-related topic for the blog post, such as "Top Destinations to Visit in 2023."

Data Gathering: The transformer-based NLP model has access to a vast amount of travel-related content from various sources. It leverages this data to understand the topic, gather information, and identify key points.

Content Generation: Using the chosen topic and the gathered information, the NLP model generates a coherent and contextually relevant article. It can create an engaging introduction, provide detailed descriptions of destinations, suggest travel tips, and even include user-generated content like reviews and testimonials.

Quality Control: While the model can generate content, it's essential to have a human editor review and fine-tune the article to ensure it meets the client's brand guidelines and quality standards. The editor can add a personal touch, correct any errors and unintended contexts, such as misinformation and bias, and make the content more engaging.

Optimization: The generated article can be optimized for SEO, with the inclusion of relevant keywords, meta tags, and headers to improve search engine rankings.

Benefits:

- **Efficiency**: Transformer-based NLP significantly reduces the time and effort required for content creation. Instead of starting from scratch, content writers can focus on editing and optimizing generated content.

- **Consistency**: The model ensures that the tone, style, and voice of the content remain consistent across multiple articles, maintaining the brand's identity.

- **Scalability**: This approach allows for the creation of a high volume of content in a relatively short time, making it suitable for businesses with demanding content needs.

- **Diverse Content**: The model can generate content on various topics, catering to a wide range of audience interests.

- **Timeliness**: It can quickly produce up-to-date content on current events, trends, or seasonal topics.

Transformer-based NLP models are valuable for content generation as they combine language understanding and creativity, making it easier for businesses to produce engaging and relevant content at scale.

Drug Interaction Prediction

Use Case: Imagine you're a pharmaceutical company focused on drug discovery, and you want to enhance the process of identifying potential drug interactions, including drug–disease and drug–drug interactions. Predicting drug interactions is crucial to ensure the safety and efficacy of new medications.

How Transformer-Based NLP Is Applied

Data Collection: You gather a large dataset of scientific articles, clinical trial reports, and medical literature related to drug interactions. This dataset contains textual information describing the effects of different drugs on the human body.

Text Mining: Transformer-based NLP models are employed to extract and analyze textual data from these sources. The models can identify mentions of drugs, their mechanisms of action, and potential interactions within the text.

Entity Recognition: NLP models recognize drug names and related entities in the text. For example, they can identify drug A and drug B in a sentence.

Contextual Understanding: Transformer models understand the context in which drug interactions are mentioned. They analyze the surrounding sentences to determine whether the interaction is positive, negative, or neutral and to what extent.

Prediction: Based on the analysis of the text, the model predicts potential drug interactions, including their severity and likelihood. For example, it may identify that drug A and drug B have a high likelihood of interacting negatively when taken together.

Validation: The predictions made by the model are validated through laboratory experiments and clinical trials to confirm the accuracy of the predictions.

Benefits:

- **Efficiency**: Transformer-based NLP accelerates the process of drug interaction prediction by automating the extraction and analysis of relevant information from a vast amount of textual data.

- **Safety**: Predicting drug interactions early in the drug discovery process helps pharmaceutical companies avoid developing medications with potential safety issues, reducing the risk to patients.

- **Cost Savings**: Identifying potential drug interactions at an early stage can save significant costs associated with clinical trials and drug development.

- **Data-Driven Insights:** NLP models provide valuable insights into the mechanisms of drug interactions and their potential impact on human health.

- **Innovation**: Leveraging transformer-based NLP allows pharmaceutical companies to stay at the forefront of innovation in drug discovery, potentially leading to the development of safer and more effective medications.

Transformer-based NLP plays a vital role in drug discovery by sifting through vast amounts of textual data to identify potential drug interactions, ultimately improving the safety and efficacy of pharmaceutical products.

Legal Document Analysis and Summarization

Use Case: Imagine you're a legal researcher working for a law firm specializing in corporate law. Your client has provided you with a massive collection of legal documents, including contracts, court cases, and regulatory filings, related to a complex merger and acquisition deal. Your task is to extract critical information from these documents efficiently and summarize them for further analysis.

How Transformer-Based NLP Is Applied

Data Ingestion: The first step is to upload all the legal documents into a digital database or repository. These documents are often in various formats, including PDFs and Word documents.

Text Extraction: Transformer-based NLP models are used to extract text from these documents. These models can accurately convert scanned images and PDFs into machine-readable text.

Entity Recognition: NLP models identify key entities, such as company names, contract terms, legal clauses, and dates, within the extracted text.

Summarization: The NLP model then generates concise summaries of the legal documents. These summaries include the main points, critical clauses, and relevant legal implications. Summarization can be either extractive (selecting and rephrasing key sentences) or abstractive (generating new sentences to capture the essence of the document).

Search and Retrieval: The legal research platform allows researchers to search for specific legal concepts, clauses, or keywords within the documents. The NLP model assists in retrieving relevant documents quickly.

Legal Analysis: Researchers can use the summaries and extracted information for legal analysis, contract comparison, due diligence, and risk assessment related to the merger and acquisition deal.

Benefits:

- **Efficiency**: Transformer-based NLP significantly reduces the time required for legal document analysis and summarization, allowing researchers to focus on higher-level legal tasks.

- **Accuracy**: NLP models can accurately extract text and recognize legal entities, minimizing the risk of overlooking critical information.

- **Consistency**: Summaries generated by NLP models are consistent and objective, reducing the potential for human bias in legal analysis.

- **Comprehensive Search**: Researchers can quickly locate relevant documents and clauses through advanced search capabilities.

- **Risk Mitigation**: Identifying critical clauses and potential legal issues early in the process helps in risk mitigation and decision-making.

- **Cost Savings**: By automating document analysis and summarization, law firms can reduce the time and costs associated with manual document review.

Transformer-based NLP is a powerful tool in legal research, enabling legal professionals to efficiently analyze and summarize vast amounts of legal documents, ultimately improving decision-making, risk assessment, and the overall quality of legal services.

Summary

In conclusion, the emergence of transformer-based NLP models, exemplified by the likes of GPT variants, has undoubtedly marked a transformative era in the field of natural language processing. These models have rewritten the rules of language understanding and generation, surpassing previous architectures like RNNs and CNNs. Their dynamic attention mechanisms and contextual word embeddings have redefined the way AI comprehends language, breaking free from fixed context windows. As they find applications in customer service, sentiment analysis, content generation, pharmaceutical research, and legal document analysis, transformer-based NLP models are at the forefront of AI advancement. Their ability to grasp contextual nuances, coupled with their adaptability and scalability, promises a future where human–computer interaction through language reaches unprecedented levels of accuracy and sophistication, paving the way for innovation across diverse industries and expanding the horizons of communication and automation.

CHAPTER 6

Exploring Chatbot Technologies

Designing a transformer-based chatbot involves a series of intricate steps, serving as the foundation for effective conversational AI. It begins with data collection and preprocessing, where the chatbot learns from conversation examples, understands user queries, and delivers relevant responses. The model employs tokenization to break down text data into manageable units. At its core, the transformer architecture, comprising encoder–decoder components, powers the chatbot.

Additional facets, including embeddings, positional encoding, self-attention, and multi-head attention, enhance the chatbot's ability to comprehend context and interdependencies within dialogues. Stacked layers facilitate intricate learning, and training fine-tunes the model's performance. Once trained, the chatbot transitions to inference, response generation, and evaluation to ensure top-tier responses. Integration into platforms and ongoing refinement based on user feedback boost its responsiveness, with the option to incorporate additional features as specific use-case demands arise.

In a comparative analysis of advanced chatbot models—GPT-4, Claude 2, and Google Bard—each brings unique strengths to the table. GPT-4 stands out for its colossal scale, multimodal versatility, enhanced fine-tuning, and efforts to mitigate biases. Claude 2 excels in maintaining conversation depth, offering customization options, and prioritizing data privacy. Google Bard, including Gemini Pro, shines with its multilingual support, harnessing Google's vast knowledge base, adept media handling, and seamless integration with Google services.

Transformer-based chatbots exhibit notable strengths, including their prowess in natural language understanding (NLU), scalability across diverse applications, and the ability to handle various data modalities. They can be fine-tuned for specific tasks and maintain context throughout interactions.

© Irena Cronin 2024
I. Cronin, *Understanding Generative AI Business Applications*, https://doi.org/10.1007/979-8-8688-0282-9_6

However, they also face challenges, such as data dependency for optimal performance, resource-intensive training, potential bias from training data, limitations in common-sense reasoning, and associated development and maintenance costs. Despite these hurdles, transformer-based chatbots hold immense promise for revolutionizing human–computer interactions, necessitating informed deployment decisions across various domains and applications.

Basic Principles of Chatbot Design

Designing a transformer-based chatbot involves key steps: data collection, tokenization, and model architecture with encoder–decoder components. Embeddings and positional encoding handle text understanding, while self-attention and multi-head attention capture context. Stacked layers enable complex learning, and training optimizes the model. Inference and response generation follow, with evaluation for performance assessment. Integration into platforms is essential, and ongoing fine-tuning based on feedback enhances responsiveness. Additional features can be added as needed for specific use cases.

A basic design of a transformer-based chatbot typically consists of several key components and explicit steps:

1. **Data Collection and Preprocessing**: Gather a dataset of conversation examples that the chatbot will learn from. This dataset should include pairs of user queries or statements and corresponding chatbot responses. Preprocess the text data by tokenizing, lowercasing, and removing any irrelevant information.

2. **Tokenization**: Convert the text data into tokens, which are individual units of text such as words or subwords. Tokenization is essential for breaking down the input and output sequences into manageable parts for the model.

3. **Model Architecture**: The core of a transformer-based chatbot is the transformer architecture. This architecture consists of encoder and decoder components. The encoder takes the user's input and encodes it into a representation that captures contextual information, while the decoder generates a response based on this representation.

4. **Embeddings**: Transform words or tokens into continuous vector representations (word embeddings) that the model can work with. Many transformer models use pretrained embeddings like Word2Vec or GloVe to initialize these embeddings.

5. **Positional Encoding**: Since transformers don't inherently understand the order of words, positional encodings are added to the embeddings to convey information about the position of words in a sequence.

6. **Self-Attention Mechanism**: The self-attention mechanism is a crucial part of transformers. It allows the model to weigh the importance of each word in the input sequence concerning all other words. This mechanism enables the model to capture dependencies and relationships effectively.

7. **Multi-Head Attention**: Transformers often employ multi-head attention, allowing the model to attend to different parts of the input sequence simultaneously. This enhances the model's ability to capture various aspects of context.

8. **Stacked Layers**: The encoder and decoder components consist of multiple stacked layers of self-attention and feedforward neural networks. These layers enable the model to learn complex patterns and representations from the data.

9. **Training**: Train the chatbot using the dataset of conversation examples. The training process involves optimizing the model's parameters to minimize a loss function, typically using techniques like backpropagation and gradient descent.

10. **Inference**: Once trained, the chatbot can be used for inference, which means it can take user queries or statements as input, encode them, and generate responses using the decoder component.

11. **Response Generation**: The response generation step involves sampling or selecting the most likely next word or token based on the model's predictions. This process may involve techniques like beam search or sampling to improve response quality.

12. **Evaluation**: Evaluate the chatbot's performance using various metrics like BLEU score, ROUGE score, or human evaluations to assess the quality of its responses.

13. **Integration**: Integrate the chatbot into the desired platform or application, whether it's a website, messaging app, or customer service portal.

14. **Fine-Tuning**: Continuously fine-tune the chatbot based on user feedback and evolving conversation patterns to improve its performance and responsiveness.

This basic design provides a foundation for building a transformer-based chatbot, and depending on the specific use case, additional features, like sentiment analysis, entity recognition, or user context tracking, can be incorporated to enhance its capabilities.

A Comparative Study: GPT-4 vs. Claude 2 vs. Google Bard

In a comparative evaluation of three advanced chatbot models, GPT-4, Claude 2, and Google Bard, distinctive features and strengths emerge. GPT-4, the latest iteration of OpenAI's GPT series, impresses with its massive scale, multimodal capabilities, improved fine-tuning, and efforts to reduce biases. Claude 2, developed by Claude.ai, excels in maintaining conversational depth, customization options, and prioritizing data privacy. Meanwhile, Google Bard, including Gemini Pro, offers multilingual support, harnesses Google's extensive knowledge base, handles rich media, and integrates seamlessly with Google services. The choice among these chatbots should be guided by specific use case requirements, where GPT-4's scale and versatility, Claude 2's depth and customization, and Google Bard's multilingual capabilities and knowledge integration cater to diverse needs.

GPT-4

GPT-4, short for "Generative Pretrained Transformer 4," is the latest iteration of

OpenAI's GPT series. It builds upon the success of its predecessors and offers several notable features:

- **Scale**: GPT-4 is known for its impressive scale, with billions of parameters, allowing it to generate highly contextually relevant responses across various topics.

- **Multimodal Capabilities**: GPT-4 can process both text and images, enabling it to understand and generate content in a more versatile manner.

- **Improved Fine-Tuning**: Fine-tuning GPT-4 for specific tasks or domains has become more effective, making it adaptable to a wide range of applications, including chatbots for customer service and content generation.

- **Reduced Bias**: Efforts have been made to reduce biases in GPT-4's responses, although challenges in this area persist.

Claude 2

Claude 2 is a chatbot developed by Claude.ai and has gained recognition for its unique features:

- **Conversational Depth**: Claude 2 is designed to engage in deep and context-aware conversations. It excels at maintaining the context of a conversation, which is especially useful for tasks requiring extended interactions.

- **Customization**: Claude 2 can be fine-tuned to specific industries or applications, making it a versatile choice for businesses with domain-specific requirements.

- **Privacy-Focused**: Claude.ai emphasizes data privacy and security, which can be a crucial consideration for organizations dealing with sensitive information.

Google Bard

Google Bard that includes Gemini Pro is Google's latest entry into the chatbot landscape and comes with its own set of strengths:

- **Multilingual Support**: Google Bard boasts impressive multilingual capabilities, making it a suitable choice for international businesses and users.

- **Knowledge Integration**: It leverages Google's vast knowledge base, enabling it to provide accurate and informative responses across a wide range of topics.

- **Rich Media Handling**: Google Bard can handle various media types, including images and audio, enhancing its ability to provide rich and multimedia-enriched responses.

- **Integration with Google Services**: It seamlessly integrates with Google services, offering convenience for users already within the Google ecosystem.

Comparative Analysis

Scale: GPT-4 has a clear advantage in terms of scale, with its massive number of parameters. This scale allows it to generate highly coherent and contextually relevant responses across a wide array of topics.

Conversational Depth: Claude 2 excels in maintaining conversational depth, making it suitable for tasks where extended interactions and context retention are crucial.

Customization: Claude 2 offers significant customization options, allowing businesses to tailor the chatbot to their specific needs.

Multilingual Support: Google Bard's multilingual capabilities make it a strong contender for businesses with a global audience.

Knowledge Integration: Google Bard benefits from Google's extensive knowledge base, making it a valuable resource for information-based tasks.

Privacy and Security: Claude 2 emphasizes data privacy and security, which can be a key consideration for organizations dealing with sensitive data.

In conclusion, the choice between GPT-4, Claude 2, and Google Bard depends on the specific requirements of the chatbot's intended use case. GPT-4's scale and versatility make it a compelling choice for many applications, while Claude 2's conversational depth and customization options cater to specific needs. Google Bard's multilingual support and knowledge integration may be ideal for businesses with a global presence. Ultimately, the selection should be based on the unique demands of the project and the desired features and capabilities.

Strengths and Weaknesses

Transformer-based chatbots, like any technological innovation, exhibit a unique blend of strengths and weaknesses that have profound implications for their application in various contexts:

Strengths

- **NLU**: At the forefront of their capabilities lies the exceptional proficiency of transformer-based chatbots in comprehending and generating natural language. These models can decipher context, nuances, and intricacies within conversations, fostering more meaningful and context-aware interactions.

- **Scalability**: Transformer-based chatbots exhibit an inherent scalability that allows them to tackle vast datasets and cater to applications of varying complexities. Their adaptability to diverse scenarios makes them versatile and suitable for a wide range of use cases.

- **Multimodal Capabilities**: A subset of transformer-based chatbots possesses the remarkable ability to process not only textual data but also other modalities, such as images, video, and audio. This multimodal prowess broadens their utility, enabling them to engage with users through diverse channels and media types.

- **Versatility**: One of the standout advantages of transformer-based chatbots is their adaptability. They can be fine-tuned to suit specific tasks, industries, or domains, ensuring that they align closely with the unique requirements of different applications.

- **Context Retention**: These chatbots are adept at retaining context throughout conversations, a vital attribute that contributes to the coherency and relevance of interactions. The ability to maintain context enables more meaningful and effective communication.

Weaknesses

- **Data Dependency**: A notable weakness lies in the data dependency of transformer-based chatbots. Achieving optimal performance necessitates substantial training data, which can be a challenge, especially in domains with limited available data.

- **Resource-Intensive**: The development and operation of large transformer models demand substantial computational resources. Training and running these models can be resource-intensive, posing cost and infrastructure challenges.

- **Potential Bias**: Transformer-based chatbots are susceptible to inheriting biases present in their training data, potentially resulting in biased or inappropriate responses. Addressing bias in AI models remains an ongoing concern in the field.

- **Lack of Common Sense**: While proficient in many language tasks, transformer-based chatbots may struggle with tasks requiring common-sense reasoning and a deep understanding of real-world contexts, which can limit their utility in certain applications.

- **Cost**: The expenses associated with the development, fine-tuning, and maintenance of transformer-based chatbots can be significant. These costs encompass not only computational resources but also ongoing monitoring and improvement efforts.

In summary, transformer-based chatbots represent a dynamic and promising technology with the capacity to revolutionize human–computer interactions. Acknowledging their strengths and weaknesses is crucial for informed decision-making when considering their implementation in various domains and applications.

Summary

To conclude, the process of creating a transformer-based chatbot is intricate and detailed, beginning with the essential steps of gathering data, segmenting it into tokens, and applying the transformer model. The core of the chatbot's learning mechanism is the acquisition of data, specifically from real interactions, encompassing a range of user inputs and corresponding replies. The process of tokenization dissects the textual content into smaller, more manageable segments. The transformer model, distinguished by its encoder and decoder structures, is fundamental to the chatbot's ability to comprehend and produce language. To better grasp the context, the chatbot utilizes embeddings and positional encodings. Embeddings are used to convert text into a continuous vector format, and positional encodings help to relay the sequence of words. These elements are crucial in allowing the model to effectively identify dependencies and relationships within the text. The addition of multiple layers further refines the chatbot's learning process, and its training is fine-tuned to optimize its efficiency.

In the phase following training, the chatbot begins to generate responses based on user inputs and is rigorously evaluated to guarantee the quality of its interactions. Its integration into diverse platforms is a significant step, and its continual refinement, informed by user feedback, improves its responsiveness. Transformer-based chatbots are adaptable, with the flexibility to incorporate specialized features tailored to specific applications, making them dynamic tools for conversation. When comparing state-of-the-art chatbot models like GPT-4, Claude 2, and Google Bard, each has unique strengths: GPT-4 stands out with its vast scale and multimodal capabilities, Claude 2 is notable for its depth and customization in conversations, and Google Bard excels in multilingual support and integrating a wide range of knowledge. These transformer-based chatbots have several advantages, such as advanced natural language understanding, scalability, and the capacity to process different types of data. Nonetheless, they face challenges like reliance on extensive data, high resource demands, potential biases, limitations in understanding common sense, and the associated costs. Despite these challenges, transformer-based chatbots have enormous potential to transform the way we interact with computers in various fields and applications, thereby significantly influencing the future trajectory of conversational AI.

CHAPTER 7

Advanced Applications of Text-Based Generative AI

The advancement of text-based Generative AI has proven to be a pivotal development in the automation of document creation and management. This technology, which encompasses the ability to generate, review, and manage documents, is now a cornerstone of efficiency within various professional sectors. Its significance is particularly noted in fields where document turnover and precision are paramount, such as in the legal, healthcare, and customer relations industries. By employing advanced Generative AI, these sectors have observed a remarkable increase in the speed and accuracy of document-related tasks, leading to more sophisticated and intelligent operational processes.

The benefits of employing Generative AI are extensive. It provides advanced template generation, allowing for the creation of dynamic, context-sensitive documents. Through the utilization of NLP, Generative AI can produce text that is remarkably human-like, suitable for intricate tasks such as legal drafting or personalized customer communication. This technology also aids in the meticulous extraction and processing of data, thus streamlining the inclusion of accurate information into documents and significantly reducing human error. In terms of sentiment analysis, Generative AI offers nuanced, accurate, and sophisticated capabilities, which are indispensable for gauging consumer sentiment in reviews, social media, and customer feedback. This allows businesses to harness deeper insights into consumer behavior and adjust strategies accordingly.

However, the application of Generative AI is not without its limitations. The technology faces challenges in ensuring consistent quality output that rivals the depth and creativity of human work. Ethical considerations around authorship and originality, biases in training data, and the environmental impact of running large AI models are ongoing concerns. Furthermore, the legal framework surrounding the ownership and

© Irena Cronin 2024
I. Cronin, *Understanding Generative AI Business Applications*, https://doi.org/10.1007/979-8-8688-0282-9_7

use of AI-generated content is still evolving, necessitating careful navigation to balance innovation with intellectual property rights and privacy laws. Despite these challenges, the continuous learning and adaptability of Generative AI systems mean that they are progressively improving, holding the promise of revolutionizing content creation and analysis even further.

Document Automation

Generative AI document automation involves using AI to generate, review, or manage documents by automating the creation of content that would typically require human input. Generative AI has a significant and evolving connection with document automation. This synergy enhances the capabilities of document automation, leading to more advanced, efficient, and intelligent processes. Here are key aspects of how Generative AI contributes to and transforms document automation:

Advanced Template Generation: Generative AI can help in creating more sophisticated templates for document automation. It can suggest or generate various sections of a document based on the context, past data, or specific requirements, making templates more dynamic and adaptable.

NLP: Generative AI, particularly through NLP technologies, can understand, interpret, and generate human-like text. This capability is crucial in document automation for generating coherent, contextually relevant content in documents, such as drafting natural-sounding language in legal documents or personalized communication in customer correspondence.

Data Extraction and Processing: Generative AI can extract relevant information from unstructured data sources, like emails or notes, and use it to populate documents. This aspect of AI helps in automating the process of gathering and inputting data into documents, reducing manual data entry and the potential for errors.

Customization and Personalization: With Generative AI, document automation systems can create highly customized and personalized documents. AI algorithms can analyze past interactions, preferences, or specific requirements of a client or situation and tailor the content of the document accordingly.

Predictive Analytics: Generative AI can predict what type of content should be included in a document based on historical data and trends. This predictive capability ensures that the automated documents are not only accurate but also contextually relevant and up-to-date.

Error Detection and Correction: AI can be employed to detect and correct errors in documents. This includes not just spelling and grammar checks but also more sophisticated checks for compliance, consistency, and factual accuracy.

Enhanced Interactivity: In more interactive document automation systems, Generative AI can enable the creation of documents that adapt in real time. For instance, as a user inputs or selects certain options, the document can automatically restructure itself to reflect those choices.

Integration with Other AI Systems: Generative AI in document automation can be integrated with other AI systems like chatbots or virtual assistants, allowing for seamless interaction where documents can be generated or modified through conversational interfaces.

Efficiency in Large-Scale Document Generation: In scenarios where large volumes of documents need to be generated, Generative AI can significantly speed up the process while ensuring each document is tailored for its specific purpose.

Continuous Learning and Improvement: Generative AI systems often have the capability to learn from each interaction and improve over time. This means that the more a document automation system is used, the better it becomes at generating effective, accurate, and relevant documents.

In summary, Generative AI dramatically enhances document automation by making it more intelligent, adaptable, and efficient. It enables the creation of highly customized, error-free documents at scale, and its integration into document automation represents a significant leap forward in how businesses and organizations manage and generate their documentation.

Case Studies

Here are several case studies that highlight the use of Generative AI in document automation across different industries:

Legal Industry: Contract Generation and Analysis

- **Firm**: A multinational law firm.
- **Challenge**: The firm faced challenges in streamlining the contract drafting process and ensuring compliance with various legal standards.

- **Solution**: The firm implemented a Generative AI system capable of automating the creation of contract templates based on predefined criteria and client information. The AI was trained on a vast repository of legal documents to understand the language and clauses typically used in contracts.

- **Outcome**: The AI system reduced the time taken to draft contracts by 50%, minimized human errors, and ensured that the documents were compliant with the latest laws and regulations.

Finance Sector: Automated Financial Reporting

- **Company**: A large investment bank.

- **Challenge**: The bank needed to automate the generation of financial reports to improve efficiency and accuracy.

- **Solution**: An AI-driven document automation tool was developed to generate financial reports by pulling data from various internal systems, analyzing it, and creating detailed, narrative reports that highlight key financial metrics and trends.

- **Outcome**: The solution enabled the bank to produce financial reports in real time, with a high level of accuracy, and allowed analysts to focus on higher-level tasks such as strategic analysis and decision-making.

Healthcare: Patient Information Management

- **Institution**: A healthcare network with multiple facilities.

- **Challenge**: The network struggled with the management of patient records and the generation of personalized patient care reports.

- **Solution**: The healthcare network implemented a Generative AI system that automated the creation of patient care documents, which included treatment plans and patient education materials, by integrating data from electronic health records.

- **Outcome**: This led to a more personalized patient experience, ensured the consistency of patient information across various facilities, and saved significant time for healthcare providers.

Marketing: Content Creation and Management

- **Agency**: A digital marketing agency.

- **Challenge**: The agency needed to produce a large volume of personalized content for various clients efficiently.

- **Solution**: By utilizing Generative AI, the agency automated the creation of content, such as blog posts, social media updates, and targeted emails, by learning the style and preferences of each client's target audience.

- **Outcome**: This not only increased content production speed but also allowed for greater personalization and improved engagement rates across marketing campaigns.

Real Estate: Property Descriptions and Listings

- **Real-Estate Company**: A national real-estate brokerage.

- **Challenge**: The company needed to create compelling and accurate property listings quickly to stay competitive.

- **Solution**: The company employed Generative AI to automate the writing of property descriptions by inputting property features, photographs, and other relevant data. The AI was trained to craft descriptions that highlight key selling points and appeal to potential buyers.

- **Outcome**: The solution expedited the listing process, improved the quality and consistency of property descriptions, and helped the company list properties faster than competitors.

Human Resources: Resume Screening and Job Descriptions

- **Corporation**: A global technology company.

- **Challenge**: The HR department was overwhelmed with the volume of resumes and the need to craft precise job descriptions for various roles.

- **Solution**: Generative AI was employed to parse and screen resumes, match qualifications with job requirements, and automate the creation of job descriptions based on role requirements and success profiles of high-performing employees.

- **Outcome**: The technology streamlined the recruitment process, enhanced the match between candidates and job requirements, and saved considerable time for HR professionals.

In each of these case studies, Generative AI document automation served to enhance efficiency, accuracy, and personalization while freeing up human resources for more complex, strategic tasks. The success of such implementations hinges on the quality of the AI training, the specificity of the use case, and the ongoing management of the AI systems to ensure they continue to learn and adapt to changing environments and requirements.

Next, another advanced application of text-based Generative AI, sentiment analysis, is discussed in terms of its tools and metrics.

Sentiment Analysis: Tools and Metrics

Generative AI in sentiment analysis represents a significant and evolving frontier in the field of AI. Sentiment analysis, also known as opinion mining or emotion AI, involves the computational task of identifying and categorizing opinions expressed in a piece of text, especially to determine whether the writer's attitude toward a particular topic, product, etc., is positive, negative, or neutral. This technology plays a crucial role in various sectors, including marketing, customer service, product analysis, and social media monitoring. The integration of Generative AI in sentiment analysis has led to more nuanced, accurate, and sophisticated analysis capabilities.

Tools for Generative AI in Sentiment Analysis

NLP Libraries

NLP libraries like NLTK, SpaCy, and Stanford NLP are foundational tools for sentiment analysis. These libraries provide essential functionalities such as tokenization (breaking text into words or phrases), part-of-speech tagging, and named entity recognition, which are crucial in preparing and understanding text data for sentiment analysis.

Machine Learning Frameworks

Machine learning frameworks such as TensorFlow, PyTorch, and Keras are used to develop and train machine learning models, including those utilized in sentiment analysis. These frameworks facilitate the construction of complex deep learning models that can process and analyze extensive text data, learning from context, syntax, and semantics to derive sentiment.

Pretrained Models

Pretrained models like GPT have been trained on vast amounts of text data and can be fine-tuned for specific tasks like sentiment analysis. These models are adept at understanding the nuances and complexities of language, making them highly effective for detailed sentiment analysis.

Cloud-Based AI Services

Cloud-based services like Google Cloud Natural Language, IBM Watson Tone Analyzer, and Amazon Comprehend offer ready-to-use sentiment analysis capabilities. These services are user-friendly and easily integrable into various applications but might offer less customization and control compared to building a bespoke model.

Open-Source Tools

Open-source tools like Hugging Face's Transformers library provide access to advanced pretrained models, such as ClinicalBERT, BioBERT, RoBERTa, T5, and XLNet. These tools are highly adaptable and can be customized to meet specific requirements of sentiment analysis tasks, offering a balance between ease of use and flexibility.

Metrics for Evaluating Sentiment Analysis

Accuracy

This metric assesses the proportion of correct predictions made by the sentiment analysis model out of total predictions. It provides a basic measure of the model's overall performance.

Precision and Recall

Precision measures the proportion of correctly identified positive results, while recall calculates the proportion of actual positives that were correctly identified. These metrics are important in scenarios where false positives and false negatives have different implications.

F1 Score

The F1 score is the harmonic mean of precision and recall, offering a single metric that balances these two aspects. It is particularly useful in situations where the class distribution is imbalanced.

Confusion Matrix

A confusion matrix is a table that is used to describe the performance of a classification model. It displays the number of correct and incorrect predictions, broken down by each class. For sentiment analysis, this matrix can reveal which specific sentiments are being confused.

ROC–AUC Score

The Receiver Operating Characteristic (ROC) curve and the Area Under the Curve (AUC) are used to assess the performance of binary classifiers. A higher AUC indicates better model performance, particularly in distinguishing between sentiment classes.

Mean Absolute Error (MAE)

In scenarios where sentiments are rated on a scale, MAE measures the average magnitude of errors in a set of predictions, disregarding the direction of these errors.

Sentiment Polarity and Subjectivity Scores

Some tools generate scores for polarity (how positive or negative the text is) and subjectivity (how subjective or objective the text is). These scores allow for a more nuanced analysis beyond simple positive/negative classification.

Challenges and Considerations

- **Contextual and Cultural Understanding**: One of the biggest challenges in sentiment analysis is the model's ability to understand context and cultural nuances. The same phrase might have different sentiments in different contexts or cultures.

- **Detecting Sarcasm and Irony**: Sarcasm and irony detection remains acomplex task for AI, as it often requires understanding beyond the literal meaning of words.

- **Data Bias and Ethical Considerations**: Ensuring that the training data for sentiment analysis is unbiased and representative is crucial. There's also an ethical consideration in how sentiment analysis is applied, especially in terms of privacy and consent.

- **Continuous Learning and Adaptation**: Sentiment analysis models, especially those powered by Generative AI, need continuous updating and learning to stay relevant. Language evolves, and new slang, phrases, or meanings emerge, which the models need to adapt to.

- **Multilingual Support**: Another challenge is developing sentiment analysis tools that work effectively across multiple languages, each with its own linguistic and cultural intricacies.

In conclusion, the integration of Generative AI into sentiment analysis has significantly enhanced the precision and complexity of our ability to assess and interpret the emotional subtext of written material. This progress marks a substantial shift in the capabilities of AI, presenting a nuanced approach to decoding human sentiments. With this advancement, industries across the board stand to gain deeper insights into consumer behavior and public opinion. However, this technological progression also brings with it the need for ongoing refinement and ethical oversight to ensure that sentiment analysis remains both accurate and fair in an ever-changing linguistic landscape. As we continue to utilize Generative AI in this capacity, it is crucial to address these challenges to fully realize the benefits of this powerful analytical tool.

Next, the overall benefits and limitations of Generative AI-driven content creation are reviewed.

Generative AI-Driven Content Creation: Benefits and Limitations

Benefits

Accelerated Production and Enhanced Efficiency

Generative AI stands out for its ability to rapidly generate content, far surpassing human capabilities in speed, which significantly optimizes content creation timelines. This swift production is indispensable in industries that demand a constant and high-volume content output like digital news outlets, content marketing firms, and social media content creation. The agility offered by AI not only accelerates production but also allows for more time to be allocated to strategic and creative endeavors, thus enhancing overall productivity.

Scalable Content Generation

One of the most striking advantages of AI in content creation is its inherent ability to scale. Unlike human-dependent processes, AI-driven content creation does not suffer from the typical constraints of increased resource needs or exponential cost hikes as production volumes grow. This scalability is particularly critical for enterprises and digital platforms where content needs are dynamic and expansive.

Personalization at Scale

Generative AI's ability to customize content based on individual user preferences or demographic data is a game-changer, particularly within the realms of marketing and advertising. The capacity to deliver personalized content at scale can dramatically bolster user engagement and, by extension, conversion rates. This level of personalization means that content can be finely tuned to resonate with diverse audiences, resulting in more effective and impactful communication.

Creative Collaboration

For creative professionals, Generative AI can act as an invaluable collaborator. It can offer novel ideas, suggest alternatives, and provide a source of inspiration, which can be particularly useful in overcoming creative blocks or expanding one's artistic horizons. This partnership can push the boundaries of creativity, offering new perspectives and approaches to traditional creative processes.

106

Breaking Language Barriers

The ability of AI to translate and localize content seamlessly is transforming global communication strategies. It ensures that content can traverse linguistic barriers, reaching wider audiences without the need for extensive human translation teams. This not only streamlines the process of making content globally accessible but also ensures consistency and accuracy across different languages and cultural contexts.

Reduction in Operational Costs

The cost-effectiveness of AI-driven content creation cannot be overstated. By automating significant portions of the content creation workflow, organizations can achieve substantial savings on operational costs. This reduction extends beyond monetary aspects, as it also minimizes the potential for human error, thereby saving additional time and resources that would otherwise be spent on corrections and quality control.

Limitations

The integration of Generative AI into content creation has catalyzed a paradigm shift in how content is conceptualized, developed, and disseminated. Yet, this innovative frontier is accompanied by a suite of intricate challenges that must be meticulously navigated to unlock the full spectrum of possibilities inherent in this technology.

Legal Limitations and Compliance

The legal landscape surrounding Generative AI is still in its infancy, and there are significant limitations and gray areas in current laws that need to be clarified. As AI-generated content becomes more prevalent, existing copyright, patent, and trademark laws may struggle to keep up with the nuances introduced by AI creations. For instance, determining the legal author of an AI-generated piece—a machine or the programmer who created the AI, or the individual who provided the initial input—presents a complex legal puzzle. Moreover, the potential for AI to generate content that infringes on existing copyrights or trademarks, either overtly or through subtle similarities, requires the establishment of new legal frameworks and guidelines to govern the creation, distribution, and ownership of AI-generated works.

These frameworks need to balance the promotion of innovation and creativity with the protection of intellectual property rights. They must address questions such as How can we ensure fair use of AI-generated content? What constitutes infringement in the context of AI? How can liability be determined and enforced? How can we protect the rights of creators and copyright holders while fostering the continued development and application of AI in content creation?

In addition, there are challenges related to data privacy and protection laws. AI systems often require large amounts of data, which may include personal information. Ensuring compliance with global data protection regulations, such as the General Data Protection Regulation (GDPR) in the European Union, adds another layer of complexity. These regulations stipulate how personal data can be collected, processed, and stored, necessitating that AI systems and their operators maintain rigorous standards of data privacy and security.

Furthermore, international legal disparities present additional hurdles. As AI-generated content can be disseminated across borders with ease, the content that is legal in one country may not be in another, leading to potential conflicts of law. International cooperation and harmonization of laws will be crucial to address these cross-border legal challenges effectively.

The legal challenges extend beyond intellectual property and privacy. There are also considerations related to consumer protection laws. For instance, if AI-generated content is misleading or incorrect, it may result in consumer harm, raising issues of liability and consumer redress. Ensuring that AI-generated content is transparently labeled and that consumers are aware of the nature of the content they are consuming is an important legal consideration.

Lastly, there is the issue of accessibility and anti-discrimination laws. As Generative AI becomes more integral to content creation, it is essential to ensure that the content is accessible to all, including people with disabilities. This means that AI-generated content must comply with accessibility standards, such as the Web Content Accessibility Guidelines (WCAG), and anti-discrimination laws to prevent the exclusion of any group from the benefits of AI advancements.

Enhancement of Quality Consistency

One of the most prominent challenges in the utilization of AI for content generation is the fluctuation in the quality of its output. While AI has demonstrated the capacity to produce content at a scale and speed unattainable by humans, the depth, authenticity, and creative essence that human creators infuse into their work can be lacking. The variance in quality ranges from highly engaging pieces that rival human work to subpar outputs that fail to resonate with audiences. This inconsistency poses a considerable barrier to the reliance on AI for quality content creation, necessitating mechanisms for quality control and standards that ensure a high caliber of AI-generated material.

Ethical and Legal Conundrums

The ascent of AI in the creative domain has sparked profound ethical and legal discussions. Issues of intellectual property rights, the definition of authorship, and the moral implications of AI-generated content sit at the heart of these debates. The line between AI-assisted content and human originality blurs, raising questions about the ownership of AI-generated works. Furthermore, there is the potential for AI to inadvertently infringe upon existing copyrights, replicate works without proper attribution, or create outputs that closely mirror pre-existing materials, potentially without intent or awareness. These issues necessitate a robust legal framework that clearly delineates the boundaries of AI in content creation and safeguards against the infringement of intellectual property.

Cultural Competence and Contextual Understanding

AI systems are often limited in their grasp of the subtleties of human context, cultural nuances, and intricate layers of social norms. This limitation can lead to content that, while factually accurate, misses the mark in terms of cultural sensitivity or appropriateness. The risk is particularly high when AI-generated content touches on sensitive social issues or cultural narratives that require a delicate, informed approach. Ensuring that AI systems are not only technically proficient but also culturally aware and contextually sensitive is a significant challenge that developers and users alike must address.

Mitigating Data Bias and Misinformation

The reliance on vast datasets for AI training means that any biases—whether intentional or inadvertent—that are present in the training material will likely be reflected in the content generated by AI. This can manifest in stereotypical portrayals, skewed narratives, or the reinforcement of prejudicial viewpoints. The propagation of misinformation is another risk, as AI may generate plausible but factually incorrect content based on flawed data inputs. It is imperative that the datasets used for training AI are diverse, balanced, and free from prejudicial biases to prevent the perpetuation of these issues.

Promoting True Creativity and Innovation

AI's proficiency in generating content that follows existing patterns is well established. However, its ability to truly innovate—to break free from the patterns it has been trained on and generate novel, original content—is still a subject of discussion. The creative process in humans is not just a matter of algorithmic pattern recognition but involves intuition, emotion, and the ability to connect disparate ideas in novel ways. Whether AI can truly emulate this aspect of human creativity or if it is intrinsically limited to iterating over learned data is a challenge that continues to fuel debate among technologists, artists, and philosophers.

Environmental Considerations

The environmental impact of AI content generation is nontrivial. The energy requirements for training and operating complex AI models are substantial, often necessitating large data centers that consume vast amounts of electricity. This energy use, often sourced from nonrenewable resources, contributes to the carbon footprint of AI technologies. Finding sustainable and energy-efficient ways to train and operate AI models is a pressing challenge, one that is critical to address in the face of global environmental concerns.

Ensuring Access and Inclusivity

The advanced technology underpinning Generative AI is not evenly distributed across the global population. There is a significant risk that these powerful tools could deepen the divide between those with access to cutting-edge technology and those without. This digital divide could lead to a concentration of power and influence among a select few, while others are left without the means to participate in this technological revolution. Ensuring that Generative AI technology is accessible and inclusive, providing opportunities for wide-ranging participation, is essential for fostering a diverse and equitable creative industry.

Summary

In summary, while Generative AI offers exciting possibilities for content creation, it also brings forth a complex array of legal, quality, ethical, contextual, bias, creativity, environmental, and accessibility challenges. Each of these challenges requires thoughtful consideration and action from multiple stakeholders to ensure that the use of Generative AI in content creation is responsible, equitable, and sustainable. The evolution of legal frameworks to accommodate and regulate AI is particularly critical and will likely be an ongoing process as technology continues to advance and its applications become more widespread.

In conclusion, the integration of text-based Generative AI into document automation and sentiment analysis heralds a new era in content management and emotional intelligence in text analysis. This technology has significantly improved efficiency and accuracy across various sectors, notably in legal, healthcare, and customer relations, by automating complex documentation processes and providing deeper insights into consumer sentiments. While it offers remarkable benefits such as sophisticated template creation, advanced data processing, and nuanced sentiment interpretation, it also confronts challenges like ensuring quality consistency, navigating ethical and legal complexities, and addressing environmental concerns. As the technology continues to evolve and adapt, its potential to revolutionize content creation and analysis, despite these hurdles, remains immense, promising a future where Generative AI-driven solutions become increasingly integral in managing and understanding the vast landscape of human-generated text.

CHAPTER 8

Sense-Based Generative AI Demystified

The realm of Generative AI typically conjures images of algorithms churning out text, images, or music. However, the burgeoning field of sense-based Generative AI represents a paradigm shift in AI, encompassing technologies that can generate sensory experiences similar to human perception. This branch of AI focuses on creating or augmenting sensory data in a way that is meaningful and interpretable to both machines and humans.

Here we explore its applications across visual, auditory, and multisensory categories and illustrate how these technologies are not only enhancing machine perception but also revolutionizing human–computer interactions.

Categories: Visual, Auditory, and Multisensory

In the innovative world of Generative AI, three key domains—visual Generative AI, auditory Generative AI, and multisensory Generative AI—stand out for their transformative impact across various industries.

Visual Generative AI focuses on creating high-resolution, detailed images using advanced techniques like GANs and transformers. This technology finds applications in diverse fields such as art creation, medical imaging, virtual and augmented reality, manufacturing, e-commerce, and geospatial imaging. GANs, with their unique structure of a generator and a discriminator, are particularly adept at producing photorealistic images. Transformers, borrowed from the success in natural language processing, have also made significant strides in handling complex visual tasks, leading to more coherent and contextually accurate visual outputs.

Auditory Generative AI delves into the realm of sound, employing technologies like RNNs, CNNs, transformers, and GANs adapted for audio processing. This branch of AI

© Irena Cronin 2024
I. Cronin, *Understanding Generative AI Business Applications*, https://doi.org/10.1007/979-8-8688-0282-9_8

has revolutionized music composition, voice synthesis, sound design for entertainment, language translation, and telecommunication enhancements. It's particularly notable for its ability to create realistic and immersive auditory experiences, from synthetic speech to complex musical compositions.

Multisensory Generative AI represents a bold leap forward, aiming to create cohesive experiences that engage multiple senses simultaneously. This field leverages a mix of machine learning frameworks, sensor technologies, robotics, and data-processing techniques to simulate human sensory experiences. Multisensory Generative AI has potential applications in healthcare, education, retail, entertainment, travel, culinary arts, art and design, and urban planning. It seeks to integrate various sensory outputs like touch, smell, and taste alongside visual and auditory elements to create fully immersive experiences.

Together, these three domains of Generative AI—visual, auditory, and multisensory—are shaping a future where AI can create experiences that are not only indistinguishable from real life but also tailored in ways that enhance and extend our human capabilities. As these technologies continue to evolve, they promise to transform how we interact with digital content, enabling new forms of creativity, innovation, and immersive experiences across multiple sectors. The following is more detail on all three areas.

Visual Generative AI

Visual AI uses techniques like GANs to create detailed and high-resolution images that can be used for everything from art creation to medical imaging. In the realm of virtual reality, visual Generative AI contributes to the development of immersive and dynamic environments that can be used for training simulations, gaming, and therapeutic purposes.

The advent of sense-based Generative AI, specifically within the visual domain, represents a monumental shift in how we create and interact with digital imagery. This shift is largely propelled by advanced machine learning models such as GANs and, more recently, transformer models which have begun to make their mark in the visual arena. These technologies are the engines behind an array of applications that extend far beyond static images, delving into dynamic, interactive, and hyper-realistic visual simulations. More detail on particular supportive technologies can be found in Chapter 9, "In-Depth Look at Supportive Visual Algorithms and Computer Vision."

Transformers in Visual AI

Transformers, known for their success in NLP, are now being adapted for visual tasks (vision transformers, ViTs). They can handle sequences of image patches and model relationships between them, making them suitable for complex tasks like image classification, object detection, and even image generation. The self-attention mechanism in transformers allows for the consideration of the entire context of an image, leading to more coherent and contextually accurate visual outputs.

GANs

GANs have been a cornerstone in visual Generative AI due to their ability to generate photorealistic images. They consist of two neural networks—the generator and the discriminator—trained simultaneously through a competitive process. The generator creates images that the discriminator then evaluates against real images, with the goal being to produce images so convincing that the discriminator cannot differentiate them from authentic photos.

Expanding Realms of Visual AI Application

Art and Creative Media:

- Artists and designers are using visual Generative AI to push the boundaries of creativity, generating novel artworks and designs. Open-source models include LLaVA 1.6, CogVLM, CogCoM, Qwen-VL, PaLI-X, PaLM-E, Fuyu-8B, AnyGPT, CoLLaVO, and CogAgent.

- Film and media production can use AI to create detailed sets, backgrounds, and visual effects, lowering costs and increasing efficiency.

Medical Imaging:

- GANs can augment medical training datasets with synthetic images that preserve patient anonymity, enhancing the training of medical professionals without compromising privacy.

- AI-generated medical imagery can assist in planning surgeries or simulating medical scenarios for educational purposes.

Virtual and Augmented Reality:

- In VR, visual Generative AI is creating environments that are richly detailed and reactive to user input, providing immersive experiences for training, gaming, or exploration.

- AR applications are benefiting from AI that can accurately overlay digital information onto the real world, enhancing user interaction with their environment.

Manufacturing and Prototyping:

- Visual AI accelerates the design process by generating multiple iterations of potential products, allowing for rapid prototyping and market testing.

- In automotive design, AI-generated visuals help in visualizing new models and features with high fidelity.

E-commerce and Retail:

- AI-generated images allow customers to see products in different colors and styles without the need for physical samples.

- Virtual try-on solutions, powered by visual Generative AI, enable customers to visualize clothing, accessories, or makeup on themselves, driving engagement and reducing returns.

Geospatial Imaging:

- AI-generated geographical and topographical maps can assist in urban planning and environmental monitoring.

- Synthetic satellite imagery can be used for strategic planning and simulation in areas where current data is unavailable or outdated.

Spotlight on Transformers

In the context of visual Generative AI, transformers represent an advanced class of neural network architectures that are particularly adept at handling sequential data, including pixels in images or sequence of images in videos. While GANs have been pivotal in generating high-quality and high-resolution visuals, transformers are beginning to play a significant role in visual tasks due to their ability to capture long-range dependencies and understand the context within visual data.

Here's an expansion on how transformers are being integrated into visual Generative AI:

Art and Design Creation: Transformers are being used to analyze and generate art by learning from vast collections of artworks. They can capture the style of specific artists or art periods and create new works that reflect those styles, offering tools for artists and designers to explore new creative territories.

Medical Imaging: In medical imaging, transformers are applied to enhance image quality and generate synthetic medical images for training and research purposes. They can learn from sequences of medical images, allowing for the creation of detailed anatomical visuals that can aid in diagnosis and treatment planning.

VR: Transformers can contribute to the creation of more coherent and contextually rich virtual environments. By understanding the sequence of user actions and visual elements, they can generate responsive environments that adapt in real time, providing more realistic training simulations for various professions, including healthcare, military, and aviation.

AR: In AR applications, transformers can help in overlaying digital information onto the real world in a way that is contextually relevant and spatially accurate. This can enhance user experiences in gaming, navigation, and information retrieval.

Video Generation and Editing: For video content, transformers can be trained to understand the temporal dynamics of video sequences, enabling the generation and editing of video content that aligns with narrative structures or follows specific directorial styles.

Autonomous Vehicles: In the autonomous vehicle space, visual transformers can process inputs from various sensors to create a comprehensive understanding of the vehicle's surroundings, improving decision-making and safety.

Fashion and Retail: Transformers are being used to generate virtual try-ons and fashion designs by understanding the context of current fashion trends and customer preferences.

Content Personalization: By understanding user preferences and content context, transformers in visual Generative AI can personalize visual content, whether it's for individual users in entertainment platforms or for targeted groups in advertising campaigns.

As transformers continue to develop, we can expect to see even more innovative applications in the field of visual Generative AI. These applications will be measured by KPIs that focus on the quality and relevance of the generated content, the efficiency of content creation, the user engagement levels, and the overall impact on the business or end-users. The integration of transformer models is poised to elevate the capabilities of visual Generative AI, creating visuals that are not just impressive in quality but also meaningful and context aware.

Challenges and Future Directions

Sense-based visual Generative AI has been advancing rapidly, creating opportunities across various sectors, from entertainment to healthcare. However, with these advancements come significant challenges and considerations for the future direction of the technology.

Challenges

Computational Resource Intensity:

High-quality visual generation requires significant computational power and memory resources, which can be costly and energy intensive.

Data Privacy and Ethics:

Training generative models often requires large datasets, which can include sensitive information. Ensuring privacy and ethical use of data is a growing concern.

Bias and Fair Representation:

AI models can inadvertently learn and perpetuate biases present in their training data, leading to unfair or stereotypical representations.

Realism vs. Uncanny Valley:

As visuals become more realistic, there is a fine line between lifelike and unsettling representations, known as the uncanny valley, particularly in humanoid figures.

Authenticity and Deepfakes:

The ability to create hyper-realistic media leads to challenges with authenticity verification, potentially giving rise to misinformation through deepfakes.

Accessibility:

Ensuring that the benefits of visual Generative AI are accessible to a broad range of users, including those in lower-income regions or with disabilities, is a challenge.

Future Directions

Advancements in Model Efficiency:

Developing more efficient algorithms and model architectures that require less computational power will make visual Generative AI more accessible and sustainable.

Federated Learning and Privacy-Preserving Techniques:

Implementing federated learning and differential privacy can help train models without compromising individual data privacy.

Bias Detection and Mitigation:

Developing tools and methodologies for detecting and mitigating biases in visual datasets and models will be crucial for fair and ethical AI.

Enhanced Realism:

Pushing the boundaries of realism while avoiding the uncanny valley will continue to be a focus, improving user experience in virtual reality and other applications.

Authentication Mechanisms:

Creating robust mechanisms to authenticate AI-generated content will become increasingly important to maintain trust and integrity in media.

Cross-Modal Integration:

Integrating visual Generative AI with other sensory AI systems (like auditory and haptic) to create cohesive multisensory experiences will be a significant step forward.

Democratization of AI Tools:

Making AI tools more user-friendly and accessible to non-experts will democratize the creation of visual content, enabling more people to harness the power of visual Generative AI.

Legal and Regulatory Frameworks:

Establishing clear legal and regulatory frameworks will be necessary to address the misuse of generative technology and protect intellectual property rights.

Ethical Guidelines and Standards:

Developing and adopting ethical guidelines and standards for the use of visual Generative AI will help guide the responsible development and deployment of the technology.

Interdisciplinary Research and Collaboration:

Encouraging collaboration across disciplines, such as AI ethics, computer science, psychology, and arts, can lead to more holistic and human-centered advancements in visual Generative AI.

The trajectory of sense-based visual Generative AI is both exciting and fraught with complexities. Addressing the challenges and moving forward with thoughtful consideration of the ethical, legal, and social implications will be essential for harnessing the full potential of this transformative technology.

Auditory Generative AI

Auditory Generative AI stands at the forefront of a revolution in sound and audio processing within the field of artificial intelligence. This specialized branch harnesses advanced technologies, including RNNs, CNNs, transformers, and GANs, to create and manipulate auditory experiences with unprecedented realism and complexity. From generating intricate musical compositions to synthesizing lifelike speech, auditory Generative AI is redefining the boundaries of how sound is produced and experienced.

This domain is integral in a variety of applications, such as enhancing musical creativity, developing realistic soundscapes in entertainment, and facilitating language translation and dubbing, among others. It also plays a vital role in creating assistive technologies and improving telecommunications. Despite its remarkable advancements, auditory Generative AI faces significant challenges, including managing authenticity and deepfakes, capturing emotional nuances in audio, addressing data privacy concerns, and overcoming computational and integration complexities. As the field progresses, it aims to enhance the emotional intelligence of systems, personalize user experiences, and develop real-time processing capabilities while maintaining ethical standards and fostering cross-disciplinary collaboration. This exciting area of AI continues to evolve, promising to expand the horizons of auditory experiences with innovative and ethically responsible technologies.

Core Technologies

RNNs:

RNNs have been pivotal in modeling time series data, such as audio signals, due to their ability to process sequences of information. They are essential in tasks where understanding the temporal context is crucial, such as in speech and music generation.

CHAPTER 8   SENSE-BASED GENERATIVE AI DEMYSTIFIED

CNNs:

While traditionally used in image processing, CNNs also contribute to audio analysis by treating spectrograms as visual inputs. This allows them to identify patterns and features within audio data effectively.

Transformers:

The self-attention mechanism in transformers, which has revolutionized NLP, is now being adapted for audio. Transformers can handle long-range dependencies in audio data, making them ideal for complex auditory tasks that require understanding over extended time periods.

GANs:

In the audio domain, GANs are used to generate high-fidelity sounds by training two neural networks in tandem: one to generate audio and the other to evaluate its authenticity. This is particularly useful in creating realistic sound effects and enhancing the quality of synthesized speech.

Expanding Applications of Auditory Generative AI

Music Composition and Production:

AI can now compose music in various genres, creating novel pieces that resonate with human emotions. It can also assist musicians by generating ideas for melodies, harmonies, and rhythms, thereby enhancing the creative process.

Synthetic Voice Generation:

Advanced voice synthesis technologies are generating voices that are increasingly lifelike, with applications in virtual assistants, audiobooks, and voiceovers. For individuals with speech impairments, such AI can create personalized synthetic voices, preserving the uniqueness of their voice.

Sound Design in Entertainment:

Generative AI is revolutionizing sound design by creating realistic and immersive soundscapes for movies, video games, and virtual reality applications. It can simulate environments, from bustling cityscapes to serene natural settings, with accurate auditory details.

Language Translation and Dubbing:

AI systems can translate spoken language and generate audio in the target language, maintaining the speaker's original intonation and style. This has significant implications for global communication and media consumption.

Assistive Technologies:

Auditory Generative AI can provide real-time descriptive audio for the visually impaired, enhancing their understanding of their surroundings or multimedia content.

Education and Training:

In educational settings, AI-generated audio can create interactive learning environments, simulate language conversations for language learning, and provide personalized feedback to learners.

Telecommunications:

AI can improve the clarity of voice calls in noisy environments by generating clean audio signals, enhancing the caller's voice, and suppressing background noise.

Challenges and Future Directions

Sense-based auditory Generative AI, which encompasses the creation and manipulation of sound using AI, is an area experiencing rapid growth and innovation. However, it presents unique challenges and directions for future development.

Challenges

Authenticity and Deepfakes:

As auditory AI becomes more advanced, distinguishing between real and AI-generated audio becomes difficult, raising concerns about the creation of convincing fake audio content.

Emotional Nuance:

Capturing the subtleties of emotion in speech and music is a complex task. AI-generated audio may still lack the depth and nuance that give human-generated sounds their richness.

Data Privacy:

Voice data is often personal and sensitive. Collecting and using such data to train AI models raises significant privacy concerns.

Computational Costs:

Generating high-quality audio in real time can be computationally expensive, requiring significant processing power and potentially limiting accessibility.

Bias and Representation:

Like all machine learning applications, auditory Generative AI can perpetuate biases present in the training data, leading to unfair or skewed outputs.

Integration with Other Systems:

Combining auditory AI with other sensory AI systems in a seamless and synchronized manner remains a technical challenge.

Future Directions

Improved Emotional Intelligence:

Future auditory AI systems will need to better understand and replicate the emotional content in human speech and music, making interactions more natural and effective.

Personalization:

Developing systems that can adapt to individual user preferences and contexts, such as accent, speech patterns, and preferred types of music.

Real-Time Processing:

Advancements in hardware and algorithms will allow for real-time generation and manipulation of audio, critical for applications in communication and entertainment.

Enhanced Security Measures:

Implementing robust security and authentication protocols to prevent misuse of voice generation and protect against voice-based security breaches.

Democratization of Content Creation:

Making auditory AI tools more accessible to creators and the public, enabling more people to produce high-quality audio content.

Multisensory Experiences:

Integrating auditory Generative AI with other sensory technologies to create immersive experiences in virtual reality, gaming, and simulation-based training.

Addressing Bias:

Developing methodologies to identify and correct biases in datasets and generated audio to ensure fair and representative outputs.

Regulation and Standards:

Establishing legal frameworks and standards to govern the use of auditory generative AI, particularly concerning copyright, deepfakes, and privacy.

Eco-Friendly AI:

Innovating more energy-efficient models to reduce the carbon footprint associated with training and running complex AI systems.

Cross-Disciplinary Collaboration:

Encouraging collaboration between AI researchers, audio engineers, psychologists, and artists to create auditory AI that is technologically advanced and culturally informed.

As auditory Generative AI continues to evolve, the focus will likely shift toward creating systems that are not only technically proficient but also ethically responsible and creatively inspiring. The future will involve a careful balance between leveraging the power of AI to expand the horizons of auditory experiences while safeguarding against the potential risks and challenges that come with such technology. Since auditory Generative AI has not been focused on enough in media and literature, more detail can be found in Chapter 10, "The Auditory and Multisensory Experience."

Multisensory Generative AI

Multisensory Generative AI represents a bold and ambitious leap toward creating machines that can interpret and synthesize stimuli across several senses simultaneously. This field of AI attempts to go beyond visual and auditory dimensions to include tactile, olfactory, and even gustatory experiences, aiming to produce a cohesive multisensory experience that mirrors human perception.

Core Technologies

Multisensory Generative AI, such as Meta's ImageBind, is an AI that aims to create outputs engaging multiple senses simultaneously and is an emerging field that leans on a variety of core technologies:

Machine Learning and Deep Learning Frameworks

Neural Networks: Multilayer perceptrons, CNNs, and RNNs are fundamental in processing and generating complex data for multiple senses.

GANs: These are particularly useful for creating realistic images and sounds by learning to mimic the distribution of real sensory data.

VAEs: Used for generating new instances of data within a specific domain, such as synthesizing new smells or tastes based on existing compounds.

Transformer Models: Adapted from NLP to process sequential data across sensory modalities, transformers can handle complex patterns in data that involve time or sequence dependencies, such as music or speech.

Sensor Technology

Electronic Sensors: Devices that can detect and measure stimuli like pressure, temperature, and chemical composition, which are essential for simulating touch and smell.

Haptic Feedback Devices: Utilized to simulate touch, including texture, vibration, and force feedback.

Electronic Nose and Tongue: Sensors capable of detecting and distinguishing odors and flavors, which can be used to generate olfactory and gustatory outputs.

Robotics and Actuators

Soft Robotics: These systems use materials and actuators that can mimic the softness and flexibility of organic tissues, providing realistic tactile feedback.

Micro-actuators: Small-scale devices that can simulate fine-grained sensory experiences, such as the feel of different materials or the movement of air to replicate a breeze.

Data Processing and Synthesis

Signal Processing: Techniques to analyze and manipulate signals for generating or altering sensory data, crucial for audio and tactile information.

Synthetic Data Generation: Algorithms to create data that can train AI systems in environments where collecting real-world data is impractical or impossible.

Virtual and Augmented Reality

VR and AR Engines: Software platforms that integrate various sensory data types to create immersive experiences.

Spatial Computing: Technology that understands the 3D space and can simulate how objects and environments should look, sound, and feel from different perspectives.

Cross-Modal AI Systems

Cross-Modal Representation Learning: AI that can learn the relationships between different modalities, such as the sound and sight of a crashing wave, to create cohesive multisensory experiences.

Multimodal Fusion Algorithms: Techniques to combine data from different sensory modalities into a single model, enhancing the AI's ability to generate complex, integrated sensory experiences.

These core technologies form the building blocks of multisensory Generative AI systems. By leveraging and integrating these various components, AI developers aim to create experiences that closely mimic the richness of human sensory perception. The continuous advancement of these technologies is critical to the development of more sophisticated and seamless multisensory AI applications.

Expanding Applications of Multisensory Generative AI

Multisensory Generative AI, with its capacity to integrate and synthesize experiences across multiple sensory domains, is rapidly expanding. This innovative field holds the promise of transforming a wide array of industries and human experiences. Here's a look at some of the burgeoning applications:

Healthcare and Therapeutics

Virtual Reality Therapy: Utilizing multisensory VR environments for treating conditions like post-traumatic stress disorder (PTSD), phobias, and anxiety disorders. These environments can simulate real-world scenarios for exposure therapy in a controlled and safe setting.

Rehabilitation: Creating simulations that help patients regain lost sensory functions, such as touch or smell, or assist in physical therapy through interactive, multisensory environments.

Medical Training: Providing medical students with a more immersive learning experience by simulating medical procedures and environments with realistic visual, auditory, and haptic feedback.

Education and Training

Immersive Learning: Enhancing educational content with multisensory inputs to provide a more engaging and effective learning experience, especially in subjects like science, history, and arts.

Special Education: Developing tailored educational tools for students with disabilities, offering sensory experiences that cater to their unique learning needs.

Professional Skill Development: Simulating real-world professional environments for training purposes, such as piloting aircraft, operating heavy machinery, or conducting scientific experiments.

Retail and E-Commerce

Virtual Try-Ons and Showrooms: Allowing customers to not only see but also feel the texture of clothes or products in a virtual setting, enhancing the online shopping experience.

Product Demonstrations: Using multisensory simulations to demonstrate product features and benefits in a more engaging and comprehensive manner.

Entertainment and Gaming

Immersive Gaming: Creating more engaging and realistic gaming experiences by incorporating tactile, olfactory, and gustatory feedback along with traditional visual and auditory stimuli.

Enhanced Media Consumption: Offering a more immersive viewing experience in films and TV by synchronizing multisensory effects with the on-screen action.

Travel and Exploration

Virtual Tourism: Enabling people to explore distant or inaccessible locations through immersive multisensory experiences, including the sights, sounds, and even the smells of the location.

Cultural Experiences: Simulating cultural events, traditions, and environments, providing users with an authentic feel of different cultures from the comfort of their homes.

Culinary Arts

Flavor and Aroma Simulation: Experimenting with flavor and aroma profiles in food and beverage development or for culinary training.

Dining Experiences: Enhancing the dining experience with synchronized visual, auditory, and olfactory stimuli, elevating the enjoyment of food.

Art and Design

Interactive Art Installations: Developing art pieces that respond to and interact with the audience across multiple senses, creating a more engaging form of expression.

Innovative Design Prototyping: Using multisensory feedback to test and refine product designs in a virtual environment, assessing aspects like ergonomics, aesthetics, and usability.

Environmental and Urban Planning

Simulating Environments: Architects and urban planners can use multisensory AI to simulate how a space will look, sound, and even feel, aiding in the design of more functional and pleasant environments.

Disaster Simulation and Planning: Creating realistic simulations of natural disasters for planning and training purposes, helping emergency responders prepare for real-life scenarios.

As technology continues to evolve, the potential applications of multisensory Generative AI are bound to expand even further, opening new frontiers in human–computer interaction. The key to unlocking its full potential lies in the continuous advancement of AI models and sensory technologies, ensuring they work in harmony to replicate the complexity of human sensory experiences.

Challenges and Future Directions

In the area of multisensory Generative AI, the integration and coordination of multiple sensory outputs present formidable challenges. These include the complexity of harmonizing different sensory data types, the demanding computational requirements for processing high-dimensional data, and the intricacies of collecting multisensory data while respecting privacy concerns. Additionally, ensuring the quality and fidelity of generated stimuli to elicit appropriate psychological and physiological responses and designing inclusive and accessible experiences for all users are critical areas of focus. Ethical considerations also play a significant role, given the potential for misuse of highly realistic simulations. Looking ahead, the field aims to develop advanced simulation technologies, tailor personalized experiences, enhance cross-modal research, integrate with virtual and augmented reality, and create sustainable, therapeutic, and quality-of-life-improving applications. As multisensory Generative AI continues to evolve, it holds the promise of revolutionizing our interaction with digital content, offering experiences that are both indistinguishably real and uniquely enhanced.

Challenges

Complexity of Integration:

Coordinating multiple sensory outputs to work in harmony is a highly complex task. Each sense involves distinct data types and processing techniques, making integration a significant challenge.

High-Dimensional Data Processing:

Processing and generating high-dimensional data across senses requires substantial computational power and advanced algorithms, which can be resource-intensive.

Data Collection and Privacy:

Gathering multisensory data, especially for personal experiences like taste and smell, is difficult and raises privacy concerns.

Quality and Fidelity:

Ensuring that the generated stimuli are realistic and immersive enough to elicit the correct psychological and physiological responses from users.

Accessibility and Inclusivity:

Designing experiences that are accessible to all users, including those with sensory impairments, is a significant challenge.

Ethical Considerations:

As with other AI technologies, there are concerns about the ethical implications of creating highly realistic simulations that could potentially be misused.

Future Directions

Advanced Simulation Technologies:

Developing new hardware and software that can simulate textures, tastes, and smells more authentically.

Personalized Experiences:

Utilizing machine learning to tailor multisensory experiences to individual preferences or therapeutic needs.

Cross-Modal Research:

Expanding research into how different senses interact with each other, known as cross-modal perception, to create more cohesive experiences.

Integration with Virtual and Augmented Reality:

Combining multisensory AI with VR and AR to create fully immersive environments for education, training, and entertainment.

Enhanced User Interfaces:

Developing user interfaces that can effectively capture and respond to multisensory user input.

Sustainable Development:

Innovating ways that ensure the sustainability of multisensory AI, minimizing the environmental impact.

Therapeutic Applications:

Leveraging multisensory AI for therapeutic purposes, such as creating environments that can help in treating PTSD, anxiety disorders, or phobias.

Improving the Quality of Life:

Using multisensory AI to enhance the quality of life for individuals with disabilities, by creating assistive technologies that cater to their unique sensory needs.

Sensory Data Synthesis:

Developing techniques to synthesize sensory data, such as digital scent and taste technologies, which can replicate real-world experiences.

Holistic Experiences in Retail:

Applying multisensory AI in retail to provide customers with a holistic experience of products before purchase, such as the feel of fabric or the scent of a perfume.

The potential of multisensory Generative AI is vast, holding the promise to revolutionize how we interact with digital content and machines. It has the potential to create experiences that are not only indistinguishable from reality but also tailored and enhanced in ways that reality cannot offer. As we venture into this largely uncharted territory, the focus will be on creating technologies that are not only advanced but also responsible and beneficial to society at large. More detail on Multisensory Generative AI can be found in Chapter 10, "The Auditory and Multisensory Experience."

Business Applications and KPIs

Sense-based Generative AI is a groundbreaking development in the field of artificial intelligence, with a wide array of business applications across various industries. The success of these applications can be measured through specific Key Performance Indicators (KPIs) tailored to each sensory category.

Visual Generative AI

Business Applications:

- **Advertising and Marketing**: Creating visually appealing ad content, personalized product previews, and dynamic social media posts.

- **Design and Architecture**: Generating architectural renderings and interior design proposals, automating the creative process.

- **E-Commerce**: Producing photorealistic images of products for online catalogs from various angles without the need for physical photoshoots.

- **Entertainment**: Crafting concept art and special effects for movies and video games, as well as full features in the future.

- **Healthcare**: Enhancing medical imaging analysis and generating synthetic data for research and training.

KPIs for Visual Generative AI:

- **Content Engagement**: Click-through rates, shares, and likes on AI-generated visual content.

- **Design Time Reduction**: Decrease in time taken from concept to final design.

- **Cost Savings**: Reduction in costs associated with traditional photography and design.

- **Accuracy**: The precision of AI-generated images compared to actual products.

- **User Experience**: Customer satisfaction scores for visual content relevance and appeal.

Auditory Generative AI

Business Applications:

- **Music Production**: Composing background scores for videos, games, or advertisements, as well as full songs and scores.

- **Customer Support**: Offering dynamic and natural-sounding voice assistants.

- **Audiobooks and Voiceovers**: Generating voice narratives for books, training materials, or explainer videos.

- **Language Learning**: Providing pronunciation guides and interactive speaking exercises with synthetic voices.

- **Alert Systems**: Creating distinct, nonintrusive auditory alerts for machinery, vehicles, or software applications.

KPIs for Auditory Generative AI:

- **Customer Interaction Rates**: Usage frequency and duration with AI-generated audio interfaces.

- **Production Efficiency**: Time and cost savings in audio content creation.

- **Quality of Experience**: User ratings for the naturalness and clarity of synthetic voices.

- **Error Rate**: The incidence of misunderstandings or miscommunications by voice AI.

- **Brand Recognition**: The effectiveness of auditory branding elements in creating recall.

Multisensory Generative AI

Business Applications:

- **VR and AR**: Developing immersive training simulations for high-risk jobs or medical procedures.

- **Retail**: Creating virtual try-on experiences that mimic the texture and feel of products.

- **Automotive**: Designing in-car experiences that integrate sight, sound, and touch for better driver engagement.

- **Real Estate**: Offering virtual tours of properties with multisensory elements to enhance the buying experience.

- **Food and Beverage**: Experimenting with flavor and aroma profiles for product development.

KPIs for Multisensory Generative AI:

- **Immersive Experience Quality**: User ratings on the realism and satisfaction of virtual experiences.

- **Training Effectiveness**: Improvement in trainee performance and reduction in on-the-job errors post-training.

- **Conversion Rates**: Increase in sales conversions from users who experienced multisensory previews.

- **Innovation Index**: Number of new experiences or products created using multisensory AI.

- **Customer Retention**: The impact of multisensory experiences on customer loyalty and repeat business.

Across all these applications, the overarching goal of sense-based Generative AI is to enhance user engagement, streamline creative processes, and deliver personalized experiences. The KPIs associated with these technologies should be continually monitored and analyzed to ensure that the AI systems are delivering on their intended value proposition and are aligned with the strategic objectives of the business.

Summary

In conclusion, it is clear that this field represents a monumental leap in artificial intelligence, transcending traditional boundaries and offering a multidimensional approach to sensory experience creation. Visual, auditory, and multisensory Generative AI are not merely technological advancements but transformative movements toward a more immersive and interactive future.

Visual Generative AI, with its prowess in creating vivid and lifelike images, is revolutionizing industries from healthcare to entertainment, offering new dimensions in visualization and design. Auditory Generative AI, on the other hand, is redefining the landscape of sound, providing groundbreaking applications in music production, language learning, and customer interaction. Multisensory Generative AI, perhaps the most ambitious of all, aims to synthesize experiences across multiple sensory domains, crafting simulations that are indistinguishable from real-life experiences.

These advancements, however, come with their own set of challenges and ethical considerations, from the complexity of integration and high-dimensional data processing to concerns about data privacy and the creation of inclusive experiences. The future of sense-based Generative AI hinges on the development of advanced simulation technologies, personalized experiences, and sustainable solutions that respect privacy and ethical standards.

The business implications of these technologies are vast, with each domain offering unique applications and measurable impacts through specific KPIs. From enhancing user engagement and creative processes in visual AI to improving customer interactions and auditory branding in auditory AI and from creating immersive training environments in multisensory AI to pioneering new product development in retail, the potential applications are as diverse as they are impactful.

Sense-based Generative AI is not just an evolution of technology; it's a new way of thinking about and interacting with the digital world. It promises a future where AI-generated experiences are seamlessly integrated into our daily lives, enhancing our perceptions and interactions in ways previously unimaginable.

In our next chapter, Chapter 9, "In-Depth Look at Supportive Visual Algorithms and Computer Vision," we detail particular supportive technologies, and in Chapter 10, "The Auditory and Multisensory Experience," we take a deeper dive into audio and multisensory Generative AI.

In-Depth Look at Supportive Visual Algorithms and Computer Vision

The integration of supportive visual algorithms and computer vision within the domain of Generative AI marks a transformative shift in how we engage with and interpret the vast quantities of visual data generated daily. In this chapter, we review the nuances and complexities of Neural Radiance Fields (NeRFs), a groundbreaking development within the sphere of computer vision, and evaluate their interplay with Generative AI technologies. NeRFs' novel approach to synthesizing three-dimensional scenes from two-dimensional imagery has carved out a pivotal role in the advancement of visual AI, enriching the capabilities of Generative AI in creating deeply immersive and realistic environments. This synergy is critical for a myriad of applications, from enhancing the authenticity of VR experiences to refining the visual fidelity of photorealistic renderings.

As we unravel the layers of these technologies, it becomes evident that NeRFs and Generative AI are not merely standalone marvels but are interconnected in a symbiotic relationship that propels both fields forward. While NeRFs focus on the precise reconstruction of 3D spaces from 2D data, their generative capabilities complement the broader creative potential of Generative AI. This partnership is vital in applications requiring nuanced visual understanding and generation, such as in AR and VR, where the convergence of accuracy and creativity is paramount.

© Irena Cronin 2024
I. Cronin, *Understanding Generative AI Business Applications*, https://doi.org/10.1007/979-8-8688-0282-9_9

Furthermore, we explore the technical and practical aspects of 3D Gaussian Splatting, another technique integral to the visualization and enhancement of three-dimensional data. Like NeRFs, this technique is instrumental in rendering continuous visual phenomena, thereby serving as a crucial component in the visualization pipeline of Generative AI outputs.

The chapter's sections provide a comprehensive overview of how these technologies not only coexist but thrive in a collaborative environment, highlighting their importance in fields like medical imaging and scientific visualization. Additionally, we will examine how the fusion of computer vision and Generative AI can be strategically employed within business contexts to foster innovation, enhance product development, and drive market research.

This exploration aims to provide a thorough understanding of these technologies, emphasizing their individual strengths and collective impact. As we review the technical intricacies and broad-ranging applications, we also address the inherent challenges and ethical considerations, underscoring the need for responsible and judicious implementation. This discourse serves as both a testament to the remarkable achievements in visual AI and a forward-looking perspective on its potential to redefine our interaction with technology and visual data.

Neural Radiance Field (NeRFs)

NeRFs represent a significant advancement in the field of computer vision and graphics. They are not a form of Generative AI in the traditional sense, like GANs or VAEs, but they do have generative capabilities and play a significant supporting role in a Generative AI pipeline, particularly in the visualization and enhancement of 3D environments. First, let's explore NeRFs in detail to understand their nature and how they function.

Understanding NeRFs

Concept: A NeRF is a framework designed to encode a 3D scene within the weights of a neural network that is fully connected, specifically within a structure known as the multilayer perceptron (MLP). It synthesizes 3D scenes from a set of 2D images and achieves this by learning a volumetric scene function. This function maps a 3D coordinate (in space) and a viewing direction to a color and a volume density.

Functionality: When a NeRF model is trained on a set of images of a scene, it learns to interpolate and reconstruct the 3D scene. This means that one can generate novel views of the scene from angles that were not in the original dataset. This is particularly powerful for applications where 3D understanding from limited 2D data is crucial.

Data-Driven: NeRFs are data-driven, meaning they rely on input data (images) to create the 3D model. The quality and diversity of the input data significantly affect the quality of the generated model.

Comparison with Generative AI

Generative Nature: While NeRFs can generate new views of a scene, they don't create entirely new content in the way that GANs or VAEs do. Generative AI typically refers to models that can produce new, unseen data (like images, text, or music) that resemble the training data but are not direct reconstructions.

Purpose and Application: NeRFs are primarily used for realistic 3D scene reconstruction and rendering, whereas traditional Generative AI models are often used for a broader range of creative and generative tasks, such as creating new artworks, generating synthetic data for training other models, or style transfer.

Training and Output: NeRFs are trained specifically on the task of understanding and reconstructing 3D space from 2D images. In contrast, generative models like GANs are trained to learn the distribution of a dataset and produce new samples from this learned distribution.

Applications of NeRFs

Photorealistic Rendering: NeRFs can create highly photorealistic renderings of scenes from sparse viewpoints.

AR and VR: In AR and VR, NeRFs can be used to create realistic environments and objects by reconstructing real-world scenes.

Film and Animation: NeRFs have potential applications in visual effects, allowing for the creation of complex 3D environments from 2D images or video footage.

Scientific Visualization: They can be used to visualize complex data in fields like medicine or astronomy.

Challenges and Limitations

Computational Intensity: NeRFs require significant computational resources for both training and inference.

Quality Dependence on Data: The quality of the reconstruction is heavily dependent on the quality and variety of the input images.

Limited Generalization: NeRFs are typically scene-specific and don't generalize well across different scenes without retraining.

While NeRFs share some characteristics with Generative AI, particularly in their ability to create new views or aspects of a scene, they are primarily focused on accurate 3D reconstruction from 2D data. They represent a fascinating blend of computer vision and graphics, with growing applications in various fields, though they come with their own set of challenges and limitations.

Even though Neural Radiance Fields (NeRFs) in themselves are not Generative AI, the significant supporting role they play in a Generative AI pipeline, particularly in the visualization and enhancement of 3D environments, warrants their discussion. Let's explore how NeRFs complement Generative AI:

Data Augmentation

Synthetic View Generation: NeRFs can be used to generate additional views of a scene from sparse data, which can be particularly useful for training other AI models that require diverse training datasets.

Detail Enhancement: By generating detailed 3D representations, NeRFs could enhance the realism of synthetic images created by generative models.

Post-processing and Refinement

Photorealism: After a generative model creates an object or a scene, NeRFs can help in rendering that scene from various viewpoints with consistent lighting and geometry, adding to the photorealism.

Fine-Tuning: Generative models might produce outputs that are not fully consistent in 3D space. NeRFs can refine these outputs to ensure they are spatially coherent from all viewing angles.

Interactive Applications

AR/VR: In AR and VR, generative models can create content, and NeRFs can help integrate this content seamlessly into real-world environments, enhancing the user's immersive experience.

Interactive Design: NeRFs could allow designers to interact with Generative AI outputs in a 3D space, making it easier to visualize and modify generated objects or environments.

Visualization of Generative Processes

3D Visualization: NeRFs can visualize the process and outcomes of generative models in 3D, which is invaluable for understanding and presenting the capabilities of Generative AI.

Data Exploration: In scientific research, generative models might be used to hypothesize data distributions or phenomena, and NeRFs could visualize these hypotheses in a more tangible form.

Hybrid Modeling

Combined Models: A pipeline could be established where generative models create base structures or textures, and NeRFs are used to project these into a consistent 3D space, creating models that are both generative and volumetrically accurate.

Theoretical Applications

Generative 3D Modeling: In theory, one could combine the principles of generative modeling with NeRFs to create a new kind of generative model that works directly in 3D space, which could revolutionize 3D content creation.

Challenges and Considerations

Computational Load: Both generative models and NeRFs are resource-intensive. Combining them would require substantial computational power.

Integration Complexity: The integration of NeRFs with generative models would require careful alignment of the data representations and objectives of both models.

Quality Control: The outputs from a combined pipeline would need to be monitored to ensure that the generative aspects and the NeRF-based rendering complement each other without introducing artifacts.

In conclusion, NeRFs could play a supportive and even transformative role in Generative AI pipelines, particularly where 3D data and environments are involved. The synergy between NeRFs and Generative AI models could lead to more realistic, detailed, and immersive experiences across various applications, from entertainment to scientific visualization.

3D Gaussian Splatting: Technical and Practical Aspects

3D Gaussian Splatting is a technique often employed in computer graphics, visualization, and volume rendering. It's a form of point-based rendering where each data point contributes to the final image through a "splat," which is typically a Gaussian function. The splatting process involves distributing the value of a data point over a region of space according to some kernel function—Gaussian being a popular choice due to its smooth falloff properties. Just like NeRFs, 3D Gaussian Splatting itself is not a form of Generative AI; however, it could play a strong supporting role in a Generative AI pipeline, particularly in the visualization and enhancement of 3D environments.

Here's a more detailed breakdown.

Definition and Process

Point-Based Rendering: In 3D graphics, rather than using polygons to represent surfaces, point-based rendering uses points in space. Each point carries information such as color, position, and possibly normal vectors.

Gaussian Function: The Gaussian function is a bell-shaped curve that is used in splatting to distribute a point's influence smoothly over its neighbors. It is defined by its mean (the center of the splat) and its variance (how wide the splat is).

Splatting Technique: Each point is "splat" onto the image plane. The contribution of each point is calculated using the Gaussian function to achieve a smooth gradient of influence across the pixels it affects. When many points are splatted onto the same region, they blend together to create a continuous surface appearance.

Volume Rendering: In the context of volume rendering, 3D Gaussian Splatting is used to project 3D scalar fields onto a 2D plane to create an image. Each data point in the volume can be thought of as a small, translucent sphere with a Gaussian density distribution.

Relation to Generative AI

As mentioned, 3D Gaussian Splatting itself is not a form of Generative AI. Generative AI typically refers to algorithms and models that learn to produce new data that is similar to a given dataset. However, there are a few points of intersection:

Data Visualization: Generative AI might use techniques like 3D Gaussian Splatting to visualize high-dimensional data in a way that's interpretable to humans.

Supporting Role: It could play a significant supporting role in a Generative AI pipeline, where AI-generated data needs to be rendered or visualized.

Hybrid Systems: There could be hybrid systems where generative models produce data that is then rendered using techniques like 3D Gaussian Splatting for visualization purposes.

Applications

Medical Imaging: It's widely used in medical imaging to visualize data from computed tomography scans or magnetic resonance imaging, allowing for the representation of soft-tissue structures.

Scientific Visualization: Helps in visualizing complex scientific data, such as fluid dynamics simulations or astronomical data.

Graphics and Gaming: Used for rendering effects like fog, smoke, or clouds, where a continuous volume needs to be represented.

Challenges and Considerations

Performance: Splatting can be computationally intensive, particularly for large datasets or high-resolution renderings.

Quality: The visual quality of the rendering can be affected by the choice of the Gaussian parameters and the resolution of the splatting.

Optimization: Various optimizations are often employed to improve the performance and quality of Gaussian Splatting, such as hierarchical splatting or adaptive splatting techniques.

In conclusion, 3D Gaussian Splatting is a valuable rendering technique in computer graphics and visualization, useful for creating smooth and continuous images from discrete data points. While it is not Generative AI, it plays a significant part in the visualization process for the output of Generative AI models.

In the next section, we look at how another supportive technology, computer vision, combined with Generative AI, can be used in business.

Computer Vision in Business Strategy

Incorporating Generative AI and computer vision into a business strategy can create a potent mix of capabilities that enable innovative solutions and services. Here's how businesses can strategize to leverage these technologies.

Understanding Computer Vision Used with Generative AI

Combining computer vision and Generative AI technologies integrates the capabilities of understanding and interpreting visual data with the power to create new data that's visually similar to the original data. Here's an in-depth look at how these two areas of AI can work together and the potential benefits and challenges they bring.

Synergy Between Computer Vision and Generative AI

Enhanced Data Interpretation: Computer vision technologies can interpret and analyze images and videos, extracting meaningful patterns, and Generative AI can then use these patterns to generate new images or augment existing ones.

Data Augmentation: Generative AI can create additional training data for computer vision models, which is particularly valuable when original datasets are limited or imbalanced.

Improved Accuracy: Generative AI can help in reconstructing missing parts of images or videos that computer vision systems rely on, potentially improving the accuracy of object detection and recognition tasks.

Creative Content Generation: Computer vision can analyze the style, structure, and content of visual data, which can then be used by Generative AI to create new, original works that maintain the essence of the source material.

Applications of Combined Technologies

VR and AR: Computer vision can track the user's interaction within a virtual environment, and Generative AI can create realistic textures and objects within these environments.

Automated Content Creation: In media and entertainment, these technologies can be used to automatically generate content such as video clips, images, and even entire scenes based on certain parameters or styles.

Medical Imaging: Computer vision can detect abnormalities in medical scans, and Generative AI can generate synthetic medical images for training radiologists without the need for patient data.

Real-Time Video Editing: Computer vision can identify objects and features in a video, and Generative AI can modify or enhance these features in real time, such as changing the weather in a live video feed.

Integrating into Business Strategy

Product Development: Use Generative AI to design new products or variations of existing products. Computer vision can aid in quality control by inspecting products on the production line.

Marketing: Generative AI can create personalized content for marketing campaigns. Computer vision can analyze consumer reactions to campaigns or products through sentiment analysis.

Customer Experience: Implement chatbots and virtual assistants powered by Generative AI for customer service. Use computer vision to enhance user interaction through gesture recognition or augmented reality features.

Supply Chain Optimization: Computer vision can monitor supply chains, track products, and optimize warehouse logistics. Generative AI can simulate supply chain disruptions to help plan for contingencies.

Market Research: Use computer vision to analyze market trends from social media and online content. Generative AI can help simulate market scenarios to predict future trends.

Personalization: Employ Generative AI to customize user experiences on websites or apps. Computer vision can tailor experiences based on user interaction patterns detected through image analysis.

Potential Applications

Retail: In-store cameras with computer vision can track inventory levels and customer behavior, while Generative AI can customize shopping experiences by generating targeted promotions in real time.

Healthcare: Use computer vision for diagnostic imaging and Generative AI for drug discovery by simulating molecular structures.

Automotive: Implement computer vision for autonomous driving systems and Generative AI for designing vehicle components.

Real Estate: Use computer vision for virtual property tours and Generative AI to design architectural variations.

Challenges to Address

Data Privacy: Ensure compliance with data protection laws when implementing systems that process personal data.

Bias and Fairness: Both computer vision and Generative AI can perpetuate biases if not carefully designed and trained on diverse datasets. Monitor for biases in AI models that could lead to unfair treatment of certain groups.

Quality Control: Ensuring the generated images are accurate representations and do not introduce artifacts is essential, especially in critical applications like medical imaging.

Integration Complexity: Merging the capabilities of both technologies requires sophisticated models and significant computational resources.

Ethical Implications: There are concerns about the potential misuse of these technologies, such as creating deepfakes or violating privacy.

Skill Gaps: Invest in training or hiring skilled personnel to develop and manage AI systems.

Strategic Planning for Implementation

Goal Setting: Define clear objectives for what the business aims to achieve with these technologies.

Resource Allocation: Allocate sufficient resources, including budget and personnel, for development and deployment.

Partnerships: Form partnerships with tech firms and academic institutions to stay at the forefront of AI advancements.

Continuous Learning: Establish a culture of continuous learning and adaptation to integrate new AI advancements into the business model.

Ethical Considerations: Develop an ethical framework for the deployment of AI to ensure responsible use.

Innovation Culture: Encourage a culture of innovation that embraces experimentation with new technologies.

Future Directions

Cross-disciplinary Research: Continued research at the intersection of computer vision and Generative AI could lead to breakthroughs in both fields.

Explainable AI (XAI): As these technologies advance, there will be a growing need for systems that can explain their decisions and outputs to users.

Regulation and Standardization: Establishing industry standards and regulatory frameworks will be crucial to ensure responsible use.

Custom Hardware Development: Advances in hardware, like AI-optimized processors, can help manage the computational demands of integrating computer vision with Generative AI.

Summary

When computer vision and Generative AI are combined, they create a powerful toolset for not only understanding and interpreting visual content but also for creating new content that can serve a variety of purposes. The integration of these technologies is poised to revolutionize fields ranging from creative industries to healthcare, but it also brings forth challenges that must be managed with thoughtful consideration of the ethical, societal, and technical implications.

In the evolving narrative of computer vision and Generative AI, the role of supportive visual algorithms such as NeRFs and 3D Gaussian Splatting stands out as a beacon of progress. These technologies embody the collaborative spirit of innovation, each playing a unique role in enhancing our ability to create and understand complex visual environments.

NeRFs have demonstrated a new way to interpret the relationship between two-dimensional images and their three-dimensional counterparts. They enable the generation of detailed, textured scenes that can be explored from various perspectives, which is an invaluable asset in fields where depth and realism are paramount. This capability is not just a technical accomplishment; it represents a meaningful advancement that broadens the horizon of what's possible in digital content creation.

3D Gaussian Splatting complements this by providing a method to render visual data with a finesse that contributes to the overall quality of the image. This technique may not be generative in the strictest sense, but its role in refining the output of generative models is indispensable, particularly in ensuring that the final visuals are as smooth and natural as possible.

The applications of these combined technologies are as diverse as they are impactful, touching areas such as VR and AR, medical imaging, and scientific visualization. They provide the tools for innovation and open up new possibilities for businesses to engage with customers and stakeholders in ways that were previously unattainable.

While the journey through this technological terrain has its challenges, the positive undertones are clear. Issues such as computational intensity and data quality requirements are being addressed through sustained research and development efforts. Moreover, the ethical application of these technologies is increasingly at the forefront of discussions, ensuring that advancements proceed with a keen awareness of their broader implications.

As we look toward the future, the integration of NeRFs, 3D Gaussian Splatting, and Generative AI models is poised to continue its trajectory of growth, underpinned by a conscientious approach to innovation. This balanced perspective acknowledges both the remarkable potential of these technologies to enhance our digital experiences and the responsibilities that come with their development.

In summary, the synthesis of computer vision and Generative AI through supportive visual algorithms like NeRFs and 3D Gaussian Splatting is a testament to the dynamic and responsible approach to technological advancement. It's a positive signal of the continuous effort to harmonize the capabilities of AI with the needs and values of society, ensuring that the benefits of these innovations are realized thoughtfully and inclusively.

CHAPTER 10

The Auditory and Multisensory Experience

The advent of Generative AI has ushered in a renaissance of sensory experience, where the confluence of sophisticated algorithms—RNNs, CNNs, transformers, and GANs—crafts a symphony of sounds that are as rich and complex as those produced by nature itself. This intricate symphony extends beyond the auditory, reaching into the realm of multisensory experiences, where the fusion of touch, sight, sound, and even scent is redefining the boundaries of what is possible.

RNNs, with their mastery over temporal sequences, capture the essence of progression in sound, lending a narrative to auditory experiences that unfold over time. CNNs dissect and discern the patterns within, translating frequencies into a visual language and back into the subtleties of sound. Transformers, with their nuanced attention mechanisms, orchestrate the interplay of audio segments, lending a dynamic range and context to the sounds they generate. GANs, the artisans of authenticity, weave soundscapes indistinguishable from reality, challenging our perceptions and inviting us to question the very nature of what we hear.

These algorithms serve as the pillars of a multisensory revolution, where AI is not only heard but also felt, seen, and smelled. They are at the heart of virtual realities that soothe the troubled mind with immersive therapy, of e-commerce platforms that simulate the tactile feedback of a garment, and of cinematic experiences that envelop the viewer in the story's very fabric. This integration promises a future where digital experiences are no longer confined to screens or speakers but envelop us in a tapestry of sensations that enrich every interaction and experience.

As we embark on this exploration of Audio Generative AI and multisensory integration, we are not just stepping into a new chapter of technology; we are entering a new dimension of human experience. A dimension where AI-generated voices inspire, where AI-composed music resonates with the depth of human emotion, and where

AI-crafted environments respond to and engage with our every sense. It's a journey into a future that promises not only to mirror the richness of the world around us but also to enhance it, push the limits of our imagination, and redefine what it means to experience and interact with sound and sensation.

Deep Dive into Algorithms Behind Sound Generation

The ingenious architects of sound—RNNs, CNNs, transformers, and GANs—are the virtuosos behind the curtain of Audio Generative AI, each playing a distinctive role in orchestrating the complex symphony of generated sound.

In the domain of Audio Generative AI, RNNs act as the maestros of memory, wielding their sequential prowess to weave the temporal tapestries of sound. Their intricate structure captures the essence of time-series data, making them indispensable in tasks where the continuity of context is paramount. From the rhythmic cadences of music to the modulations of speech, RNNs trace the lineage of auditory sequences with precision.

CNNs, with their adept feature extraction capabilities, transform audio into a visual spectacle of spectrograms, decoding the intricate patterns embedded within. These networks, akin to artists, paint the auditory landscape, distinguishing the subtlest of nuances in sound and enabling machines to classify and generate rich, textured audio with the discernment of a skilled musician.

Transformers, the grand strategists of attention, dissect and understand the complex layers of audio data. With the finesse of a composer, they harmonize the disparate segments of sound, creating a coherent narrative that flows through time and crafting audio experiences that are both intricate and expansive.

GANs, the alchemists of authenticity, conjure the indistinguishable from the real. In their dance of deception, the generator and discriminator push each other toward excellence, synthesizing sounds that blur the lines between artificiality and reality, challenging our perceptions, and expanding the horizons of auditory experience.

More detail on these algorithms used in Audio Generative AI is as follows.

RNNs

Recurrent Neural Networks (RNNs) are foundational to Audio Generative AI due to their architecture, which is designed to handle sequential data—data where the order and context matter. In the domain of audio, this is particularly relevant as sound is inherently a time-series signal where each moment is dependent on what came before.

Structure and Functioning of RNNs in Audio

Sequence Modeling: RNNs are adept at modeling sequences due to their recursive structure which processes one input at a time while maintaining a "memory" (hidden state) of what has been computed so far. This hidden state acts as a form of short-term memory, carrying information from previously processed inputs along the sequence.

Temporal Dependencies: In audio processing, RNNs excel at capturing temporal dependencies. For instance, in speech recognition, the way a word is pronounced can be influenced by the previous word, and RNNs can model this contextual information. Similarly, in music generation, the choice of the next note often depends on the preceding sequence of notes, and RNNs can capture these musical patterns.

Backpropagation Through Time (BPTT): RNNs are trained using a method called Backpropagation Through Time, where the network is unrolled across time steps, and gradients are calculated for each time step. This allows the network to update its weights not only based on the current input but also on the sequence of inputs it has processed.

Variants of RNNs in Audio Applications

Long Short-Term Memory (LSTM): LSTMs are a special kind of RNN capable of learning long-term dependencies. They have a complex gating mechanism that controls the flow of information, allowing them to remember important inputs over long sequences and forget the irrelevant ones. This is particularly useful in audio applications where certain sounds or patterns are critical for the overall structure, such as in the case of melody or rhythm in music.

Gated Recurrent Units (GRUs): GRUs are another variant of RNNs that simplify the gating mechanism used by LSTMs while still effectively capturing temporal dependencies. They are used for similar applications as LSTMs but are often faster to train due to their simpler structure.

Challenges with RNNs

Vanishing Gradient Problem: RNNs are notorious for the vanishing gradient problem, where gradients become smaller and smaller as they are propagated back through each time step during training, leading to difficulty in learning long-range dependencies.

Computational Intensity: Processing long sequences with RNNs can be computationally intensive, as each step depends on the previous step, making parallelization challenging.

Overfitting on Training Data: RNNs, with their significant capacity for memorization, can overfit to the training data, capturing noise rather than the underlying patterns.

Future of RNNs in Audio Generative AI

Despite these challenges, RNNs remain a powerful tool for Audio Generative AI. Future developments are likely to focus on optimizing these networks to handle longer sequences more efficiently, possibly by integrating attention mechanisms that allow the network to focus on relevant parts of the input sequence. Additionally, there is ongoing research into developing more sophisticated RNN architectures that can better balance memory capacity and computational efficiency.

In practice, RNNs have enabled breakthroughs in various audio applications, such as generating music that adapts to a listener's mood, creating dynamic sound effects for video games that respond to on-screen action, and developing more natural-sounding voice assistants capable of understanding and generating human-like speech. As the field of Audio Generative AI continues to evolve, RNNs will undoubtedly play a central role in shaping the auditory landscapes of the future.

CNNs

Convolutional Neural Networks (CNNs) have garnered attention in the field of audio processing, particularly in the generation and analysis of sound, by leveraging their strong pattern recognition capabilities. Here's a detailed look at how CNNs are applied within Audio Generative AI:

Spectrogram Analysis with CNNs

Spectrogram Transformation: A spectrogram is a visual representation of the spectrum of frequencies of a sound signal as they vary with time. By converting audio signals into spectrograms, CNNs can treat audio as a two-dimensional "image" where one axis represents time and the other frequency, and the intensity of colors represents the energy or loudness at each frequency at each point in time.

Feature Extraction: CNNs consist of layers of convolutions that can extract hierarchical features from spectrograms. The first layer may detect simple features such as edges, which in the context of audio, correspond to the onset of sounds or changes in pitch. Deeper layers can identify more complex patterns that can represent specific characteristics of the sound, such as timbre or rhythm.

Temporal Convolutional Networks: For audio applications, Temporal Convolutional Networks (TCNs), a type of CNN, can be particularly effective. They use 1D convolutions across the time dimension to capture temporal dependencies and can be used for tasks like audio synthesis and source separation.

CNN Applications in Audio

Sound Classification: CNNs can classify sounds within an audio clip, such as identifying different types of environmental sounds (e.g., rain, traffic, birdsong) or recognizing various musical instruments.

Speech Recognition: In speech recognition, CNNs can detect phonemes and other speech features from spectrograms, which can then be used to transcribe spoken words.

Music Generation: In the realm of music generation, CNNs can analyze patterns in spectrograms to generate new pieces of music by learning from the structure and style of existing compositions.

Audio Enhancement and Restoration: CNNs can also be used to enhance audio quality by identifying and isolating noise patterns in spectrograms and then subtracting these from the original signal to clean up the audio.

Training CNNs for Audio Tasks

Data Augmentation: To train CNNs for audio tasks, data augmentation is often used to increase the diversity of the training data, which can include pitch shifting, time stretching, and adding background noise to sound samples.

Transfer Learning: Leveraging pretrained models on image datasets is a common practice. These models can be fine-tuned on spectrogram data, allowing CNN to utilize learned image feature extraction capabilities for audio tasks.

Challenges and Considerations

Computational Complexity: Training CNNs on spectrograms can be computationally demanding due to the high dimensionality of audio data, especially for high-fidelity sounds.

Loss of Phase Information: Spectrograms contain information about frequency and amplitude but lose phase information, which can be important for accurately reproducing the sound.

Real-Time Processing: Using CNNs in real-time audio applications presents challenges due to the need for rapid processing and inference, requiring optimization and efficient model architectures.

As CNNs continue to be a staple in Audio Generative AI, ongoing research is addressing these challenges, enhancing the capabilities of CNNs to deal with complex audio synthesis, recognition, and enhancement tasks. With further advancements, CNNs hold the potential to drive significant progress in creating and understanding audio content with a level of detail and precision that closely mirrors human auditory perception.

Transformers

Transformers have emerged as a powerful architecture in Audio Generative AI, extending their influence beyond their initial success in NLP. The key feature that makes transformers especially suitable for audio applications is their self-attention mechanism, which dynamically weighs the importance of different parts of the input data. Let's delve into how transformers are being adapted for audio and their potential within the field.

Core Mechanisms of Transformers in Audio

Self-attention Mechanism: This allows transformers to focus on different parts of the audio signal, determining how each segment of the audio should influence the others. For example, in a conversation, the meaning of a sentence can be affected by the words that precede or follow it, and transformers can capture these dependencies regardless of the distance between relevant audio segments.

Positional Encoding: Since transformers do not inherently process sequential data in order, positional encodings are added to the input embeddings to give the model information about the order of the sequence. In audio processing, this means that transformers can understand the temporal sequence of sounds, which is crucial for tasks that require timing, such as beat detection in music or speech rhythm analysis.

Transformer Applications in Audio Generative AI

Audio Synthesis: Transformers can be trained to generate high-quality audio waveforms. They can learn from a dataset of audio samples to produce new sounds that match the style and complexity of the training data.

Music Generation: With their ability to handle long-range dependencies, transformers are ideal for generating music that has a coherent structure throughout. They can model the relationship between different musical elements across a piece, creating compositions with a sense of progression and thematic development.

Voice Separation and Enhancement: Transformers can separate individual voices from a noisy environment by focusing on the frequency and time domains where the speech signal is dominant. They can also enhance speech by attending to parts of the audio that contain speech and filtering out the noise.

Sound Event Detection and Classification: By analyzing the context of the audio signal, transformers can detect and classify specific sound events, such as the breaking of glass or the honking of a car, even when these events occur in a complex acoustic environment.

Training and Optimization for Audio

Efficient Attention: Audio signals are typically sampled at a high frequency, leading to long sequences for transformers to process. Efficient attention mechanisms, such as local attention or sparse attention, have been developed to reduce the computational burden.

Spectral Representations: Just as with CNNs, transformers can operate on spectral representations of audio-like spectrograms or mel-frequency cepstrum coefficients (MFCCs), which provide a compact representation of the audio signal.

Multi-head Attention: This feature of transformers involves running several attention mechanisms in parallel, which allows the model to focus on different types of information simultaneously, such as pitch, timbre, and rhythm in a piece of music.

Challenges and Future Directions

Resource Intensity: The self-attention mechanism of transformers is computationally intensive, especially for long audio sequences, which necessitates optimization for real-time processing and generation.

Adapting to Audio Modality: While transformers have been successful in NLP, adapting them to audio requires consideration of the unique properties of sound, such as its continuous nature and the importance of phase information.

Integrating Temporal and Frequency Information: Future work in transformers for audio involves better integration of both time and frequency information to improve tasks like audio super-resolution or full-bandwidth synthesis.

Transformers in Audio Generative AI represent an exciting frontier that is still being explored. Their flexibility and power make them ideal for a range of audio applications, from creating new sounds and music to enhancing and understanding existing audio. As research progresses, we can expect transformers to become increasingly central to the field of Audio Generative AI, driving innovations that could transform how we interact with sound in digital environments.

GANs

GANs have found a significant place in Audio Generative AI due to their unique structure and capability to generate new data instances that are indistinguishable from real data. In audio applications, GANs are used for a variety of tasks such as generating music, synthesizing speech, and enhancing sound quality. Here's a deeper look into how GANs function in the realm of audio.

Structure of GANs in Audio

Generator and Discriminator: A GAN consists of two neural networks competing against each other. The generator network creates synthetic audio, while the discriminator network tries to distinguish between the synthetic audio and real audio samples. This adversarial process encourages the generator to produce more realistic audio over time.

Latent Space: The generator network in a GAN takes a random noise vector from a latent space as input and generates audio data. The latent space offers a compact representation from which complex audio patterns can emerge as the generator learns during training.

GAN Applications in Audio Generative AI

Music Generation: GANs can be trained on a dataset of music from a particular genre or style and can then generate new music that mimics that style. The ability of GANs to capture and replicate the probability distribution of the training data makes them well-suited for this task.

Speech Synthesis: GANs are used to synthesize speech that sounds like a human voice. They can learn from a dataset of spoken words and generate new words or sentences that retain the speaker's voice characteristics.

Audio Super-Resolution: In audio super-resolution, GANs can enhance the resolution of audio signals. They take a low-resolution audio input and generate a high-resolution version by learning from examples of high-resolution audio.

Sound Effect Generation: GANs can generate realistic sound effects for use in video games, movies, and virtual reality. They can create sounds like footsteps, rain, or engine noise that can be used to enhance the auditory experience of a scene.

Training GANs for Audio

Adversarial Training: The training of GANs is a minimax game where the generator tries to maximize the probability of the discriminator making a mistake, while the discriminator tries to minimize this probability. This adversarial training helps in producing audio that is increasingly realistic as the generator learns from the feedback of the discriminator.

Feature Matching: To help stabilize the training of GANs, feature matching techniques can be employed where the generator tries to match the discriminator's internal representation of real data. This helps in avoiding mode collapse, a common issue where the generator produces a limited varieties of output.

Conditioning: GANs can be conditioned on additional information, such as class labels or even other modalities of data, to direct the audio generation process. For instance, a GAN could be conditioned on a visual input to generate sound effects that match a particular scene.

Challenges with GANs in Audio

Training Stability: GANs are known for being difficult to train due to issues like mode collapse and non-convergence. Researchers have proposed various techniques to mitigate these issues, such as using different loss functions, normalization techniques, and architectural changes.

Evaluation Metrics: Evaluating the quality of generated audio can be challenging since there are no clear and objective measures for the realism and diversity of audio like there are for images. Human evaluation is often required, which can be subjective and time-consuming.

Temporal Coherence: Maintaining temporal coherence in generated audio is crucial, especially for longer audio samples. This is an area where GANs can struggle, as ensuring that the generated audio makes sense over time is not a trivial task.

As GANs continue to evolve, their potential in Audio Generative AI grows. They are becoming more sophisticated with the integration of additional neural network components and training techniques, which opens up new possibilities for realistic and complex audio generation tasks. Whether it's for entertainment, communication, or creative expression, GANs hold the promise of significantly enriching the auditory experiences that AI can offer.

In synthesizing the intricate symphony of algorithms that give rise to the art of sound generation in AI, we have traversed a landscape rich with computational innovation and creativity. RNNs have laid the temporal groundwork, allowing us to encode and regenerate the nuanced sequences of audio. CNNs have lent their pattern-recognizing prowess to interpret the frequencies and rhythms hidden within sound waves. Transformers, with their self-attention mechanisms, have taught us to understand the context and the complex interplay of sounds over extended periods. GANs have pushed the boundaries of what is possible, challenging us to discern the created from the natural in the auditory realm.

As we look toward the horizon of Audio Generative AI, we see a future vibrant with potential—where the voices of AI not only mimic but inspire, where the music of AI resonates with the personal touch of human creativity, and where the sounds of AI enrich our lives in ways yet to be imagined. This conclusion is not an end but a gateway to countless beginnings, as each algorithm continues to learn, adapt, and evolve, playing its part in the grand orchestra of AI.

Multisensory Integration: Case Studies and Applications

Multisensory Generative AI is revolutionizing various sectors by creating immersive, interactive experiences that engage multiple senses. Let's delve deeper into specific applications and case studies in healthcare and therapeutics, retail and e-commerce, and entertainment that illustrate the transformative impact of this technology.

Healthcare and Therapeutics

Enhanced Virtual Reality Therapy for PTSD: Advanced multisensory VR environments are being used to treat PTSD patients more effectively. These environments incorporate synchronized visual, auditory, and olfactory stimuli to recreate scenarios that patients can safely navigate. The controlled settings allow therapists to gradually expose patients to traumatic memories, helping them process and manage their responses in a therapeutic context.

The intersection of healthcare, therapeutics, and advanced technologies like multisensory Generative AI is forging new pathways in the treatment of post-traumatic stress disorder (PTSD). This condition, characterized by severe anxiety, flashbacks, and uncontrollable thoughts about a traumatic event, can be debilitating. However, through the integration of multisensory virtual reality (VR) environments, new therapeutic approaches are emerging that show promise in aiding patients to cope with and recover from PTSD.

Multisensory VR Environments in PTSD Therapy

Immersion Therapy: VR technology immerses patients in a controlled, virtual world where they can be exposed to trauma-related cues without real-world risks. This exposure therapy is designed to help patients confront and reprocess the traumatic event within a safe environment.

Sensory Integration: Multisensory environments incorporate visual, auditory, and sometimes olfactory stimuli to create a comprehensive scenario that closely mimics the context of the traumatic event. By synchronizing these stimuli, the VR experience becomes more convincing, helping patients to engage with the therapy more effectively.

Progressive Exposure: Therapy typically follows a graded approach, where patients are initially exposed to less challenging aspects of the traumatic memory and gradually progress to more difficult ones. This can help prevent overwhelming the patient and allows for better management of anxiety.

Interactivity: Patients can interact with the virtual environment, often using VR controllers or body movements. This active participation can increase their sense of control and agency, which is often compromised in PTSD.

Biofeedback Integration: Some VR systems are equipped with biofeedback mechanisms that monitor physiological responses like the heart rate and galvanic skin response. Therapists can use this information to adjust the level of exposure in real time, tailoring the therapy to the patient's needs.

Clinical Evidence and Research

A growing body of research indicates that VR exposure therapy can be effective for treating PTSD. Studies show that VR-based treatments can lead to significant reductions in PTSD symptoms, often with longer-lasting effects compared to traditional exposure therapies. Research also suggests that incorporating multiple senses into the therapy may enhance the therapeutic process, potentially leading to better outcomes.

Technological Developments

Customizable Scenarios: Advances in AI and VR have led to the development of highly customizable scenarios. Therapists can design specific environments tailored to individual patient experiences, improving the relevance and effectiveness of the therapy.

Real-Time Adaptation: AI algorithms can analyze patient responses in real time and adapt the scenario accordingly. This ensures that the therapy remains within therapeutic thresholds, optimizing safety and efficacy.

Haptic Feedback: The use of haptic devices in VR therapy for PTSD can simulate touch and physical sensations, adding another layer of immersion. For instance, vibrations can simulate the feeling of an explosion or the rumble of a vehicle, depending on the traumatic event.

Challenges and Considerations

Accessibility: While promising, multisensory VR therapy is not yet widely available and can be expensive. Efforts are underway to make this technology more accessible to clinics and patients.

Individual Differences: Not all patients may respond well to VR therapy, and some may experience cybersickness or discomfort in VR environments. Personalization and patient screening are essential to ensure suitability for the therapy.

Ethical and Privacy Concerns: The use of such immersive technology raises questions about patient privacy and the ethical use of recreating traumatic scenarios. Clear guidelines and patient consent are crucial.

Long-Term Efficacy: More research is needed to understand the long-term efficacy of multisensory VR therapy for PTSD and how it compares to traditional methods over extended periods.

The use of multisensory VR in the treatment of PTSD represents a significant step forward in the application of technology in mental health therapy. By leveraging the immersive and interactive capabilities of VR, coupled with the nuanced understanding of AI, therapists can offer a powerful tool to help individuals process and overcome the challenges associated with traumatic memories. As research progresses and technology becomes more accessible, multisensory VR therapy has the potential to transform the landscape of therapeutic options available to PTSD sufferers.

Retail and E-Commerce

In the retail and e-commerce sectors, multisensory Generative AI is revolutionizing the way customers interact with products online by simulating the in-store experience. Advanced Virtual Try-On Systems, empowered by multisensory AI, are becoming increasingly sophisticated, offering a richly interactive and immersive shopping experience that extends beyond the visual element to include touch and sound.

Advanced Virtual Try-On Systems

Integration of Haptic Technology: Haptic feedback technology integrates tactile sensations into the virtual try-on experience. For example, when a customer tries on a watch virtually, they can feel vibrations or pressure on their wrist that simulate the weight and texture of the watch. This sensory feedback is achieved through wearable haptic devices or specialized interfaces that can replicate the sensation of different materials and weights.

3D Visualization and Interaction: AI-driven algorithms generate 3D models of products that customers can interact with. Users can rotate, flip, or zoom in on the product to view it from different angles, just as they would examine it in a physical store. The realistic visualization is often rendered in real time, adjusting to the movements and choices of the user.

Auditory Feedback: Adding to the sensory experience, auditory feedback provides realistic sounds associated with the products. For the virtual watch example, this could include the ticking of the watch or the sound of a clasp clicking into place. These sound effects are generated by AI to match the actions taken by the user, such as "clicking" a button or "zipping" a virtual garment.

Personalization and AI Recommendations: Multisensory AI systems can personalize the shopping experience by learning from the user's interactions and preferences. They can recommend products that match the user's style or previous purchases and adjust the virtual try-on experience accordingly. For instance, if a user prefers leather watches, the system might simulate the specific texture of leather more prominently.

Applications in Retail and E-Commerce

Clothing and Apparel: Customers can try on clothes virtually, feeling the texture of the fabric through haptic feedback devices. This can include the sensation of smooth silk, rough denim, or the weight of heavy wool.

Jewelry and Accessories: Shoppers can experience the heft and finish of jewelry items, such as the coolness of metal or the roughness of gemstones.

Furniture and Home Goods: Consumers can explore the texture and material composition of furniture, experiencing the softness of cushions or the grain of wood finishes.

Footwear: Virtual try-ons for shoes can simulate fit and comfort levels, including the stiffness of new shoes or the cushioning of insoles.

Technological Enhancements

Machine Learning for Texture Simulation: Machine learning algorithms analyze high-resolution images and physical properties of materials to accurately simulate textures in a virtual environment.

AR Fittings: Using AR technology, customers can see how products look on them or in their homes through their smartphone cameras, complementing the haptic and auditory feedback for a cohesive experience.

Real-Time Feedback Loop: The AI system can adjust the virtual experience in real time based on user feedback, improving the accuracy of simulations and enhancing customer satisfaction.

Challenges and Future Directions

Hardware Limitations: The need for additional hardware for a full haptic experience might limit accessibility for some users.

Standardization: There is a need for standardizing the sensory feedback so that experiences are consistent across different platforms and devices.

User Experience (UX) Design: Designing intuitive and user-friendly interfaces for multisensory experiences remains a challenge, as it requires seamless integration of visual, auditory, and tactile elements.

Data Privacy and Security: Collecting and processing user data to personalize experiences raise concerns about data privacy and security that need to be addressed.

In summary, multisensory Generative AI in retail and e-commerce is creating a paradigm shift in how consumers shop online, providing an experience that is closer to the physical act of shopping. As this technology matures, it has the potential to significantly enhance customer engagement, reduce return rates by providing more information pre-purchase, and open up new avenues for interactive marketing and customer experience innovation.

Entertainment

In the entertainment industry, multisensory cinema experiences are redefining the boundaries of traditional movie-watching by creating an environment where audiences can engage with a film through more than just sight and sound. This multisensory integration aims to enhance the emotional and physical connection to the on-screen narrative, making the viewer an active participant in the cinematic journey.

Multisensory Cinema Experiences

Sensory Effects Synchronization: Theaters are incorporating specialized equipment that synchronizes physical effects with specific moments in the film. Seats may move or vibrate during action sequences, fans generate wind to simulate stormy conditions, and heat lamps intensify scenes with fiery visuals.

Olfactory Immersion: Advanced scent-dispersion systems release carefully curated fragrances that complement the visuals. For example, the scent of rain may permeate the theater during a storm scene, or the smell of flowers may accompany a garden scene, providing a deeper level of immersion.

Tactile Feedback: Equipped with haptic feedback technology, theater seats can provide a tactile experience that mirrors the action on screen. This could be as subtle as the brush of air to mimic a passing object or as intense as a jolt during an explosion.

Future: The AR Holodeck Concept: Drawing inspiration from science fiction, the concept of an AR Holodeck represents the next frontier in entertainment. It refers to a fully interactive and immersive environment where virtual elements are overlaid onto the physical space, allowing individuals to interact with the storyline in real time.

Technological Enhancements for Entertainment

Interactive Storytelling: Multisensory cinema opens up new possibilities for interactive storytelling, where the audience's physical responses could influence the narrative's direction, akin to a choose-your-own-adventure style but in a cinematic format.

Wearable Technology: The use of wearable tech in theaters could provide personalized sensory feedback, such as wristbands that pulse with the movie's soundtrack or jackets that mimic the sensation of touch or temperature changes.

Spatial Audio Systems: Advanced audio systems that provide 3D spatial sound can accurately position sounds in the theater, giving the impression that audio sources move independently of the screen, enveloping the audience in the film's soundscape.

Intelligent Lighting: Dynamic lighting systems that adjust the theater's ambiance to reflect the mood and tone of the film, providing a more nuanced visual backdrop that complements the on-screen action.

Applications in Entertainment

Cinema: Beyond traditional theaters, specialized venues offering multisensory experiences could become popular destinations, especially for blockbuster releases that seek to provide an unparalleled viewing experience.

Theme Parks: Theme parks can incorporate multisensory theaters as attractions, offering short films with high-impact sensory effects that complement the park's overall theme.

Exhibitions and Installations: Art and cultural exhibitions could use multisensory experiences to bring stories and historical events to life, offering an educational experience that is both informative and emotionally engaging.

Challenges and Future Directions

Cost and Accessibility: The cost of equipping theaters with the necessary technology can be high, potentially limiting the accessibility of multisensory cinemas to a wider audience.

Content Creation: Filmmakers and studios will need to consider the multisensory experience during the production process, which may require new techniques and collaboration with technology providers.

Standardization: Creating a standardized set of practices for multisensory content can ensure a consistent quality of experience across different venues.

Research and Development: Continued R&D is essential to refine the technologies involved, reduce costs, and expand the range of sensory effects that can be accurately simulated.

The potential of multisensory cinema experiences in entertainment is vast, with current applications only scratching the surface of possibilities. As technology advances, we may see the development of fully interactive cinema environments that blur the line between film and reality, providing audiences with experiences that are more vivid and engaging than ever before. The concept of an AR Holodeck could eventually become a reality, offering a new realm of storytelling that could fundamentally transform the entertainment landscape.

Summary

In conclusion, the algorithms that drive Audio Generative AI—RNNs, CNNs, transformers, and GANs—have each contributed a unique verse to the composition of sound, bringing us closer to replicating the richness and complexity of human auditory experiences.

RNNs have demonstrated their prowess in capturing the sequential and temporal essence of sound, making them essential for tasks that require an understanding of the progression of audio over time. CNNs, on the other hand, have shown their strength in extracting features from audio, treating sound as a visual spectacle to be analyzed and understood. Transformers have brought a new depth to audio processing, with their ability to focus on different parts of an audio signal, determining the influence of each segment over others. Meanwhile, GANs have astounded us with their capacity to create audio so authentic it challenges our perception of what is real and what is generated.

The convergence of these technologies in multisensory integration has opened up a realm of applications that once seemed like the stuff of science fiction. From the therapeutic use of multisensory VR to help PTSD patients process traumatic memories, to advanced virtual try-on systems in e-commerce that emulate the tactile experience of shopping, and to multisensory cinemas that envelop viewers in a fully immersive narrative, the boundaries continue to be pushed further.

However, this is not a final act but rather an intermission. The journey into the future of Audio Generative AI and multisensory experiences is just beginning. As these technologies evolve, they promise to not only enhance our sensory experiences but to transform them, creating new ways for us to interact with the world and with each other. The algorithms will continue to learn and adapt, playing their part in the grand orchestra of AI, and leading us to a future where AI-generated voices inspire, AI-composed music resonates with the human touch, and AI-crafted sounds enrich our lives in ways we have yet to fully comprehend.

The horizon of Audio Generative AI is vibrant with potential, beckoning us to a future where the digital and sensory realms are seamlessly intertwined. It is a future that holds the promise of revolutionizing entertainment, therapy, retail, and beyond. As we step forward, we carry with us the knowledge that the auditory and multisensory experience, as crafted by AI, is not just about simulating reality—it's about enhancing it, adding new dimensions to our perception, and ultimately, enriching the human experience.

CHAPTER 11

Autonomous AI Agents: Decision-Making, Data, and Algorithms

Autonomous AI Agents, including those that are used in multi-agent systems and that use Generative AI, are decision-making systems that can independently create new content, solutions, or data that mimic real-world patterns or distributions. Since Generative AI encompasses a set of ML algorithmic techniques that allow the model to generate new instances of data that can pass for real data, it's a step beyond predictive analytics. While predictive models interpret data, generative models produce data. These agents are designed to perform tasks that require the creation of novel outputs based on learned patterns and contexts. Both this chapter and Chapter 12, "Text-Based Generative Intelligent Agents: Beyond Traditional Chatbots and Virtual Assistants," are concerned with Autonomous AI Agents. In the present chapter, we first review Autonomous AI Agents in general and then those that use Generative AI, as well as key ML algorithms that Autonomous AI Agents use. Next, we discuss data analytics and their importance and how the combination of ML and data analytics is significant for Autonomous AI Agents. In Chapter 12, "Text-Based Generative Intelligent Agents: Beyond Traditional Chatbots and Virtual Assistants," we delve into those Autonomous AI Agents that are text-based and use Generative AI which are called Text-Based Generative Intelligent Agents (GIAs).

© Irena Cronin 2024
I. Cronin, *Understanding Generative AI Business Applications*, https://doi.org/10.1007/979-8-8688-0282-9_11

Key Characteristics and Functionalities of Autonomous AI Agents

Autonomous AI Agents as systems or software that can perform tasks or functions on their own, without human intervention, in a variety of environments and situations are designed to make decisions and take actions based on their programming and the data they receive from their surroundings. The "autonomy" in these systems refers to their ability to operate independently for extended periods, adapting to changes and learning from new experiences. These Autonomous AI Agents can operate across a number of modalities such as text, audio, and video and can also utilize the Internet as needed.

Here are some key characteristics and functionalities of Autonomous AI Agents:

- **Perception**: They have the ability to perceive their environment, often through sensors or data inputs, which can include visual, auditory, or other sensory data.

- **Decision-Making**: Autonomous agents can process the data they collect to make informed decisions. They use algorithms that can evaluate multiple options and choose actions that align with their goals or objectives.

- **Learning**: Many autonomous agents are equipped with ML capabilities, allowing them to learn from their experiences and improve their performance over time.

- **Adaptation**: They can adapt their behavior in response to changing conditions in their environment, ensuring their actions remain effective and relevant.

- **Action**: These agents can take physical actions (such as a robot moving items in a warehouse) or digital actions (like a software program automatically adjusting a digital marketing campaign).

- **Autonomy**: They can function without direct human control, making them useful for tasks that are repetitive, dangerous, or require operation in remote or inaccessible locations.

- **Interaction**: Some autonomous agents can interact with humans or other systems, understanding and responding to voice commands, gestures, or other forms of communication.

- **Goal-Oriented**: They are typically designed with specific goals in mind, whether it's completing a particular task, maximizing efficiency, or learning a new skill.

Examples of Autonomous AI Agents include the following:

- **Self-driving Cars**: These vehicles combine sensors and ML to navigate roads and traffic without human drivers.

- **Drones**: Unmanned aerial vehicles that can perform tasks like surveillance, delivery, or environmental monitoring.

- **Robotic Vacuum Cleaners**: These devices can navigate home spaces, avoiding obstacles while cleaning floors.

- **Chatbots and Virtual Assistants:** Software programs that can understand and respond to human language to provide information or assistance.

- **Text-Based Generative Intelligent Agents**: Text-Based Generative Intelligent Agents go steps further than chatbots and virtual assistants, in that from an initial set of instructions they autonomously act on those instructions and, once completed, deliver the solutions.

- **Automated Trading Systems**: Programs that can execute stock trades based on market data without human intervention.

Autonomous AI Agents are becoming increasingly sophisticated and are used in a wide range of industries, from manufacturing and logistics to healthcare and entertainment. Their ability to operate independently, learn from their surroundings, and make decisions makes them a powerful tool for innovation and efficiency in the digital age.

Key Aspects of Autonomous AI Agents Using Generative AI

Autonomous AI Agents are advanced systems with the capability to extend the boundaries of innovation across various domains. These agents are not just consumers of information but also prolific producers, capable of creating and extrapolating data to foster growth and development in their respective fields.

Data Generation: Such agents can generate new data points, simulate scenarios, or create content that is similar to a given dataset but does not replicate it. This is often used in data augmentation, where new data points are needed to train other AI models.

Creativity: They can produce creative works like art, music, literature, or design elements by learning from existing styles and generating new creations that reflect learned patterns.

Problem-Solving: These agents can generate a variety of solutions to complex problems, enabling them to propose multiple viable options in situations like engineering design or strategic planning.

Prediction and Simulation: They can simulate possible future events or scenarios, which are particularly useful for risk assessment, strategic planning, and training simulations.

Adaptive Learning: Generative AI agents can adapt to new data inputs and evolve their generative processes, producing increasingly refined outputs over time.

Examples of Autonomous AI Agents Using Generative AI

Autonomous AI Agents are revolutionizing the way we handle data-sensitive and creative tasks, leveraging their ability to generate new, valuable content while adhering to ethical standards. From healthcare to gaming, these agents provide innovative solutions, ensuring privacy, enhancing user experience, and driving forward industries like design, robotics, and pharmaceuticals with unprecedented efficiency.

Synthetic Data Creation: In fields like healthcare, where privacy is a concern, autonomous agents can generate synthetic patient data for research and training, preserving individual privacy while providing valuable datasets.

Procedural Content Generation: In video games, these agents can create infinite landscapes, levels, or characters that keep the game fresh and engaging.

Design and Engineering: Generative AI can propose multiple design options for products or architectural projects, which can then be refined by human professionals.

Robotics: Robots in manufacturing might use generative models to troubleshoot and propose optimizations for production processes or to adaptively learn how to handle new objects.

Drug Discovery: AI agents can generate molecular structures that may lead to new pharmaceuticals, expediting the drug discovery process.

Natural Language Generation: Agents can write articles, reports, or code, given a set of parameters or initial inputs, using models like GPT.

Importance of Generative AI in Autonomous Agents

Generative AI stands at the forefront of technological advancement, driving autonomous agents toward new horizons of creativity and problem-solving. These agents embody a transformative power, delivering tailored, scalable solutions across industries, thus reshaping the landscape of innovation and efficiency.

Innovation: They foster creativity and innovation, breaking free from the limitations of existing data or designs.

Efficiency: They improve efficiency, generating solutions faster than traditional R&D processes.

Scalability: They provide scalability, capable of generating vast amounts of content without additional human labor.

Customization: They enable high levels of customization, as they can generate unique outputs tailored to specific requirements or preferences.

Training and Development: They are invaluable for creating realistic training environments in simulators for various professional fields.

As Generative AI continues to advance, the capabilities of these autonomous agents will expand, leading to more sophisticated applications that can act independently across various industries and domains.

Key ML Algorithms for Autonomous AI Agents

Autonomous AI Agents are systems designed to operate independently, using machine learning algorithms to perceive, decide, and learn in dynamic environments. These algorithms form the core of Generative Autonomous AI Agents, enabling them to process complex data and adapt over time.

Supervised Learning Algorithms

Linear Regression: Predicts outcomes based on input variables.

Logistic Regression: Estimates probabilities for classification tasks.

Support Vector Machines (SVM): Separates classes using hyperplanes.

Decision Trees: Models decisions in a tree-like structure.

Random Forest: An ensemble method using multiple decision trees.

Gradient Boosting Machines (GBM): Optimizes models additively.

Unsupervised Learning Algorithms

K-Means Clustering: Groups data into distinct clusters.
   Principal Component Analysis (PCA): Reduces data dimensionality.
   Autoencoders: Learns efficient codings in an unsupervised manner.
   Hierarchical Clustering: Creates a hierarchy of clusters.

Semi-supervised Learning Algorithms

Label Propagation: Combines labeled and unlabeled data for learning.
   Self-training: Enhances supervised algorithms with unlabeled data.

Reinforcement Learning Algorithms

Q-Learning: Learns action values without a model.
   Deep Q Network (DQN): Applies deep learning to approximate Q-values.
   Policy Gradients: Learns a direct mapping from states to actions.
   Actor-Critic Methods: Integrates value and policy-based methods.
   Proximal Policy Optimization (PPO): Balances exploration with exploitation.

Deep Learning Algorithms

Convolutional Neural Networks (CNNs): Ideal for grid-like data such as images.
   Recurrent Neural Networks (RNNs): Models sequential data.
   Long Short-Term Memory Networks (LSTMs): RNNs that learn long-term dependencies.
   Generative Adversarial Networks (GANs): Two networks trained in tandem for generation tasks.
   Transformers: Leverages self-attention mechanisms for handling sequence data, excelling in tasks requiring an understanding of context and sequence relationships, such as language translation and text summarization.

Evolutionary Algorithms

Genetic Algorithms: Mimics natural selection to solve problems.
   Evolution Strategies: Optimizes problems in continuous domains.

Applications in Autonomous AI Agents

Navigation and Pathfinding: Reinforcement learning, like DQN and PPO, for navigating environments.

Perception and Object Recognition: CNNs for image-based tasks.

Decision-Making: Decision trees and random forests for rule-based decisions.

Anomaly Detection: SVMs and autoencoders for identifying unusual patterns.

Sequence Prediction and NLP: RNNs, LSTMs, and transformers for language understanding and generation.

A Note on Transformers in Autonomous AI Agents

Transformers significantly enhance the capabilities of Autonomous AI Agents, particularly in interpreting and generating sequences. Their self-attention mechanisms allow for a more sophisticated understanding of temporal and contextual data, vital for agents operating in complex, time-sensitive environments. This makes transformers indispensable for tasks like real-time decision-making, natural language understanding in human–robot interactions, and processing sensor data streams for environmental perception.

The incorporation of transformers into the suite of ML algorithms for Autonomous AI Agents marks a step forward in their evolution, providing advanced capabilities in pattern recognition, decision-making, and adaptation. These agents are set to become increasingly sophisticated, with the potential to autonomously perform a wide range of tasks in unpredictable and complex environments.

Data Analytics Techniques and Their Importance

Autonomous AI Agents as systems equipped with the capability to perform tasks without human intervention rely on various data analytics techniques to understand their environment, make decisions, and learn from experiences. These data analytics techniques are primarily used to process, analyze, and interpret data, enabling these agents to make informed decisions and learn from their environment. The following is an outline of these techniques and their importance.

Descriptive Analytics

Overview: This involves summarizing historical data to understand changes over time.

   Techniques:

- **Data Aggregation and Statistics**: Summarizing data features like mean, median, mode, and standard deviation.

- **Data Visualization**: Using charts, graphs, and heatmaps for a visual summary of data trends.

   Application in AI Agents: Helps in understanding the historical performance and operational patterns.

Diagnostic Analytics

Overview: Focuses on examining data to understand the causes of past events and behaviors.

   Techniques:

- **Correlation Analysis**: Identifying relationships between different data variables.

- **Root Cause Analysis**: Determining the primary cause of a problem or event.

   Application in AI Agents: Useful for troubleshooting issues or understanding the agent's past actions.

Predictive Analytics

Overview: Utilizes statistical techniques to forecast future events.

   Techniques:

- **Regression Analysis**: Predicting a numerical value based on historical data trends.

- **Time Series Analysis**: Analyzing time-ordered data points to forecast future values.

   Application in AI Agents: Essential for anticipating future scenarios and preparing responses.

Prescriptive Analytics

Overview: Involves the use of optimization and simulation algorithms to advise on possible outcomes.

Techniques:

- **Optimization Models**: Finding the best solution from a range of possible options.

- **Simulation Techniques**: Creating and analyzing a digital simulation of a real-world process.

Application in AI Agents: Helps in decision-making by considering various alternatives and their potential impacts.

Exploratory Data Analysis (EDA)

Overview: An approach to analyzing datasets to summarize their main characteristics, often using visual methods.

Techniques:

- **Histograms, Box Plots**: For distribution analysis

- **Scatter Plots**: To find patterns, relationships, or anomalies

Application in AI Agents: Enables discovery of patterns, anomalies, or features not evident with standard analysis methods.

Data Mining

Overview: The process of discovering patterns in large datasets involving methods at the intersection of ML, statistics, and database systems.

Techniques:

- **Association Rule Learning**: Discovering interesting relations between variables in large databases

- **Cluster Analysis**: Grouping a set of objects in a way that objects in the same group are more similar to each other

Application in AI Agents: Useful for uncovering hidden patterns and structures in data, which can guide decision-making.

Sentiment Analysis

Overview: Often used to gauge public opinion, sentiment analysis involves analyzing text data to understand the sentiment behind it.

Application in AI Agents: Particularly relevant for agents interacting with humans, like chatbots or customer service bots, to understand and respond to human emotions.

Network Analysis

Overview: Involves analyzing complex networks and understanding relationships and flows between network nodes.

Techniques:

- **Social Network Analysis**: Understanding social structures through nodes (individual actors, people, or things) and ties (relationships or interactions).

Application in AI Agents: Useful in scenarios where AI agents need to understand and navigate complex systems like logistics networks, social media networks, etc.

Each of these techniques plays a crucial role in enhancing the capabilities of Autonomous AI Agents, enabling them to process and analyze data more effectively, and thereby perform their tasks with greater efficiency and accuracy.

Importance of Data Analytics Techniques in Autonomous AI Agents

Data analytics stands as the cornerstone of modern Autonomous AI Agents, equipping them with the adaptability to evolve and the acumen for informed decision-making. These sophisticated techniques not only streamline operations by enhancing efficiency and safety but also extend the reach of AI, scaling to complexities beyond human handling and crafting personalized user interactions.

Adaptability: Data analytics enable autonomous agents to adapt their behavior based on the analysis of incoming data streams, improving performance over time.

Decision-Making: Through the analysis of large datasets, agents can make informed and autonomous decisions in real time.

Situational Awareness: Agents use data analytics to understand and interpret their environment, which is vital for navigation and task execution.

Efficiency: Predictive analytics and optimization contribute to the efficient operation of autonomous systems, reducing waste and improving outcomes.

Safety: By analyzing sensor data, agents can detect potential hazards and avoid accidents, which is especially important for autonomous vehicles.

Scalability: Data analytics techniques enable agents to handle complex tasks that would be unmanageable for humans, such as monitoring extensive networks or managing large-scale automation systems.

Personalization: For agents interacting with users, such as virtual assistants, analytics allow for personalized experiences based on user data and preferences.

The data analytics techniques employed by Autonomous AI Agents are crucial for their functionality. They enable these agents to process and understand vast amounts of data, make autonomous decisions, and interact with both their environment and human users effectively. As technology advances, the role of data analytics in the development of Autonomous AI Agents will continue to grow, leading to more sophisticated, reliable, and versatile systems.

Combining ML and Data Analytics for Optimal Results

ML and Data Analytics are two pillars of modern AI systems, and when combined, they can significantly enhance the performance of Generative Autonomous AI Agents. Here's how the integration of these disciplines can yield optimal results:

Integration of ML and Data Analytics for Autonomous AI Agents

Data Acquisition and Preprocessing:

- **Data Analytics**: Involves the extraction, cleaning, and normalization of data from various sources, which may include sensors, logs, user interactions, and more.

- **Machine Learning**: Relies on this prepared data to train predictive models or to extract patterns.

- **Optimal Result**: Clean and well-structured data improves the accuracy and efficiency of ML models, leading to more reliable autonomous decisions.

Feature Engineering and Selection:

- **Data Analytics**: Identifies which aspects of the data (features) are most relevant to the problem at hand through exploratory data analysis.

- **Machine Learning**: Utilizes these features to develop models that can predict, classify, or take decisions autonomously.

- **Optimal Result**: Selecting the right features ensures that ML algorithms can focus on the most informative data, reducing complexity and improving performance.

Model Training and Validation:

- **Data Analytics**: Helps in the design of experiments and the evaluation of model performance against historical data.

- **Machine Learning**: Algorithms learn from data to create a model that can generalize from past experiences to new data.

- **Optimal Result:** Robust models that are well-validated can perform consistently in real-world conditions.

Predictive Analytics:

- **Data Analytics**: Uses statistical techniques to interpret data patterns and predict future events.

- **Machine Learning**: Enhances predictive analytics by providing models that can adapt and improve over time.

- **Optimal Result**: Accurate predictions enable AI agents to anticipate future states or events and take proactive measures.

Real-Time Decision-Making:

- **Data Analytics**: Provides real-time insights and situational awareness.

- **Machine Learning**: Employs these insights to make or recommend decisions autonomously.

- **Optimal Result**: AI agents can make informed decisions on the fly, essential for applications like autonomous vehicles or dynamic routing.

Feedback Loops and Reinforcement Learning:

- **Data Analytics**: Measures the outcomes of decisions and actions taken by the AI agent.

- **Machine Learning**: Uses feedback to reinforce successful behaviors and correct unsuccessful ones.

- **Optimal Result**: Continuous improvement in the agent's performance, as it learns from its environment and experience.

Anomaly Detection and Safety:

- **Data Analytics**: Detects deviations from normal behavior, which could indicate a problem.

- **Machine Learning**: Processes these anomalies to improve system safety and reliability.

- **Optimal Result**: Early detection of potential failures or safety hazards, allowing for preventive measures.

Scalability and Efficiency:

- **Data Analytics**: Ensures that insights are scalable and actionable across large datasets and multiple agents.

- **Machine Learning**: Provides algorithms that can operate efficiently at scale.

- **Optimal Result**: AI agents that can scale up their operations without a loss in performance, crucial for widespread deployment.

Personalization and User Experience:

- **Data Analytics**: Understands user behavior and preferences.

- **Machine Learning**: Adapts the agent's behavior to individual users' needs.

- **Optimal Result**: Personalized experiences that increase user satisfaction and engagement.

Continuous Learning and Adaptation:

- **Data Analytics**: Tracks changes in the environment or in the data over time.

- **Machine Learning**: Adjusts models in light of new data, ensuring that agents remain effective as conditions change.

- **Optimal Result**: AI agents remain relevant and effective over time, even as their operating conditions evolve.

By combining ML and Data Analytics, Autonomous AI Agents can achieve enhanced cognition and decision-making capabilities. This synergy is vital for developing systems that can operate autonomously in complex, dynamic environments, providing optimal performance and adaptability. The collaboration between the two fields creates a continuous cycle of learning and improvement, driving the evolution of intelligent systems.

Summary

Generative AI is revolutionizing the field of autonomous decision-making, serving as the robust scaffold upon which AI agents construct intricate frameworks of innovation and creation. These systems, designed for autonomy, are not merely passive recipients of pre-existing data; they are active participants in the data lifecycle, contributing original, contextually relevant data where none may have previously existed. They transcend traditional roles not only by predicting outcomes based on existing patterns but also by crafting entirely new paradigms that mimic the nuanced complexity of real-world data. This transformative approach allows for the expansion of capabilities in sensitive areas such as healthcare, where synthetic data creation upholds privacy while still providing the rich datasets necessary for advanced research and development.

The essence of Autonomous AI Agents lies in their multifaceted functionalities. Perceiving and interacting with their environment through advanced sensorial inputs, they process vast amounts of data, making informed decisions autonomously. Their ML cores are imbued with the capacity for adaptation and self-improvement, enabling these agents to evolve their decision-making prowess and enhance their efficiency in real-time scenarios. They navigate physical and digital realms with a degree of precision that mirrors human intuition but is powered by algorithms and data-driven insights, thus providing solutions that are both ingenious and pragmatic.

Moreover, these agents are equipped with the remarkable ability to generate content, simulate complex scenarios, and propose innovative solutions across various domains. In the world of video games, for example, they are the architects of endless digital terrains and dynamic characters, rendering each gaming experience unique and perpetually engaging. In the domain of design and engineering, their proposed blueprints for products or infrastructure synthesize aesthetics with functionality, offering human professionals a multitude of perspectives from which to refine final designs.

In the manufacturing sector, robots infused with generative AI capabilities are not only performing tasks but are also reimagining production processes, suggesting enhancements, and learning to manipulate novel objects with a dexterity that rivals human workers. The pharmaceutical industry is witnessing a seismic shift as AI agents accelerate drug discovery by generating potential molecular structures, thus reducing the timescales traditionally associated with bringing new medicines to market.

Natural language generation, another forte of these agents, showcases their ability to produce written content, ranging from analytical reports to creative literature, with only minimal initial input. This capability has profound implications for content creation, programming, and even journalism, as it allows for the rapid synthesis of information into coherent, contextually relevant narratives.

The ingenuity of Generative AI within autonomous agents is matched only by its efficiency. By automating the generation of solutions, these agents offer a speed of development that outpaces traditional research and development processes. The scalability of their solutions is equally significant; they possess the capability to generate vast amounts of content without a proportional increase in human labor, thus redefining productivity. This scalability extends into customization, as generative models are adept at crafting outputs that are tailored to the intricate and varied specifications dictated by different users or applications.

Training and development scenarios, particularly those that leverage simulation technologies, benefit immensely from these agents. Realistic environments, replete with challenges and variables, are conjured to provide professionals with experiential learning opportunities that are both safe and instructive. The breadth of application for these agents thus spans the gamut from practical to creative endeavors, marking them as pivotal to the ongoing evolution of numerous industries.

As we look toward the future, the trajectory of Generative AI suggests a landscape where the boundary between human ingenuity and algorithmic creativity becomes increasingly blurred. The sophisticated applications of these autonomous agents, which can now act independently across various sectors, are only beginning to unfold. Their potential to autonomously conceive, design, and execute tasks heralds a new era of intelligent systems that are not just tools but collaborators in the creative and problem-solving processes. This partnership between human and machine intelligence, facilitated by the intricate dance of data and algorithms, promises a future replete with possibilities that are as vast as they are exhilarating.

CHAPTER 12

Text-Based Generative Intelligent Agents: Beyond Traditional Chatbots and Virtual Assistants

Generative Intelligent Agents (GIAs) and Generative AI are deeply intertwined concepts within the realm of artificial intelligence. Generative AI refers to the subset of AI technologies that can autonomously generate complex content, such as text, images, audio, and other media that resemble human-like creations. GIAs are a specific application of Generative AI, embodying systems or agents that utilize these generative capabilities to perform tasks, solve problems, and interact with their environment in an intelligent manner.

Key Features of Text-Based GIAs

Text-based GIAs are a subset of AI systems that specialize in understanding, processing, and generating human language in the textual form. Leveraging advanced techniques in NLP and ML, these agents can compose and produce text that is coherent, contextually relevant, and often indistinguishable from text authored by humans. Some features of text-based GIAs are as follows:

Language Generation: They can autonomously generate written content such as stories, reports, code, and even poetry. This is made possible through models like GPT (Generative Pretrained Transformer), which can predict and generate sequences of text based on the input it receives.

© Irena Cronin 2024
I. Cronin, *Understanding Generative AI Business Applications*, https://doi.org/10.1007/979-8-8688-0282-9_12

Contextual Understanding: Text-based GIAs are designed to grasp the subtleties of language, including slang, jargon, and idioms, across various contexts. They can maintain the thread of a conversation or narrative, ensuring that their generated text aligns with the preceding content.

Interactivity: These agents can engage in conversations with users, dynamically generating responses in real time. They are often employed in generative, nontraditional chatbots, customer service assistants, and virtual companions.

Personalization: By analyzing user data and previous interactions, text-based GIAs can tailor their language and content style to suit the preferences of individual users, enhancing the user experience.

Adaptive Learning: As they interact and receive feedback, text-based GIAs can learn and adapt their language models to improve their performance in generating text.

Content Scalability: Text-based GIAs can produce large volumes of text quickly and efficiently, making them valuable for content creation at scale.

Applications of Text-Based GIAs

Automated Journalism: Generating news articles and summaries based on data inputs

Creative Writing: Composing creative pieces like fiction, poetry, or scripts

Business Intelligence: Creating business reports, summaries, and analyses from large datasets

Education: Providing personalized learning materials and interactive tutoring

Programming: Writing and optimizing code snippets or even whole programs

Customer Service: Powering generative, nontraditional chatbots and virtual assistants that handle inquiries and provide support

Challenges with Text-Based GIAs

Ethical Concerns: Issues related to the generation of biased, offensive, or inaccurate content.

Authenticity: Difficulty in distinguishing between human-generated and AI-generated text, which can lead to misinformation.

Dependence on Data: The quality of output is highly dependent on the quality and breadth of the training data.

Text-based GIAs are transforming the way we interact with and leverage written content across various domains. As their capabilities continue to advance, they open up new possibilities for efficiency and creativity while also presenting unique challenges that must be addressed to ensure their responsible use.

Comparative Analysis: GIAs vs. Traditional Chatbots and Virtual Assistants

In the rapidly evolving landscape of AI, GIAs and traditional chatbots and virtual assistants represent two distinct approaches to AI-driven interaction and task execution. Understanding the differences between these technologies is crucial for comprehending their potential applications, limitations, and impacts on various industries and user experiences.

Interaction and Response Capabilities

Traditional Chatbots and Virtual Assistants: These systems are typically programmed with a set of predefined responses. They operate based on rule-based logic or simple machine learning algorithms, which allow them to handle straightforward tasks or answer specific queries based on keywords or set patterns.

GIAs: In contrast, GIAs use advanced machine learning models, like GPT-4, to generate responses and content. This allows for more dynamic interaction, as GIAs can produce novel answers, simulate conversation, and even create content that wasn't explicitly programmed into their systems.

Learning and Adaptation

Traditional Chatbots and Virtual Assistants: Their learning capacity is generally limited to the data they were trained on. They may struggle with unexpected queries or nuances in language and often require manual updates or retraining to improve their capabilities.

GIAs: GIAs are designed to learn continuously from interactions, adapting their responses over time. They can understand context, remember past interactions, and refine their outputs, making them more sophisticated with each interaction.

Creative and Generative Abilities

Traditional Chatbots and Virtual Assistants: They are not inherently creative. Their responses and content generation are bound by their programming, making them less flexible in generating new content or ideas.

GIAs: The hallmark of GIAs is their ability to generate original content. Whether it's writing a poem, composing music, or coming up with new product ideas, GIAs have a creative edge that allows them to produce outputs that can be indistinguishable from human-generated content.

Personalization and Contextual Awareness

Traditional Chatbots and Virtual Assistants: They offer a degree of personalization, typically through user data analysis. However, their ability to contextualize responses based on past interactions or external data is limited.

GIAs: They excel in personalization and contextual awareness. By analyzing and learning from user interactions and external data sources, GIAs can tailor their responses and content to the individual user, enhancing the user experience significantly.

Application Scope

Traditional Chatbots and Virtual Assistants: They are highly effective in structured environments with specific tasks, such as customer service inquiries, basic data retrieval, and routine task execution.

GIAs: GIAs have a broader application scope due to their generative and adaptive capabilities. They can be used in creative fields, complex problem-solving, personalized tutoring, and anywhere where novel content generation is valuable.

While traditional chatbots and virtual assistants remain valuable for specific, routine tasks, GIAs represent a leap forward in AI's interactive and creative capabilities. The generative nature of GIAs opens up new possibilities in AI-human interaction, making them suitable for a wider range of applications that require flexibility, learning, and innovation. As AI continues to evolve, the distinction between these two types of systems will become increasingly significant, shaping the way AI is integrated into various sectors and everyday life.

Examples of Complex Text-Based GIAs

Complex GIAs have found applications in various fields, each exploiting their unique capabilities to analyze, predict, generate, and assist in decision-making processes. The following are some examples illustrating their diverse uses:

Business Strategy and Market Analysis Tools: GIAs in the business sector analyze market trends, consumer behavior, and competitor strategies. They generate comprehensive reports and strategic insights, helping companies to identify new market opportunities, optimize marketing strategies, and make informed business decisions. For instance, a GIA could analyze social media trends and news articles to advise on product development or marketing strategies.

Financial Analysis and Trading Algorithms: In finance, GIAs are employed to read through and interpret vast amounts of financial news, reports, and market data, generating predictions and insights for investment strategies. They can assist traders by providing real-time analysis and recommendations on stock or asset trading.

Legal Research Assistants: Legal GIAs are designed to sift through case laws, legal journals, and documents to assist in legal research. They can draft legal documents, predict litigation outcomes based on historical data, and provide insights into complex legal scenarios.

Healthcare Data Analysis Systems: In healthcare, GIAs analyze medical research, patient data, and clinical trial reports to assist in diagnosis, treatment planning, and research. They can generate patient-specific reports and assist in identifying potential treatment paths based on historical data.

Educational Tutors and Content Creators: GIAs in education provide personalized tutoring, generate educational content, and assist in curriculum development. They can analyze educational materials and student performance data to create customized learning plans and materials.

Public Policy Analysis Tools: These GIAs analyze public data, policy documents, and social trends to assist in policymaking. They can simulate the impacts of policy decisions, generate policy recommendations, and analyze public sentiment on various issues.

Interactive Storytelling and Content Creation: In the entertainment industry, GIAs are used for generating scripts, stories, and even interactive narrative experiences. They can analyze user preferences and cultural trends to create engaging and personalized content.

Crisis Management and Response Systems: Employed in emergency and crisis situations, these GIAs analyze real-time data from news, social media, and official reports to provide insights into crisis situations, helping in effective response planning and resource allocation.

Research and Development Tools: GIAs aid in R&D by analyzing scientific literature, patents, and research data, generating hypotheses, or suggesting areas for innovation. They can accelerate the research process by identifying potential breakthroughs or unexplored areas.

Customer Service and Support Bots: Advanced customer service GIAs go beyond basic chatbots by providing in-depth analysis of customer queries, generating personalized responses, and handling complex customer service scenarios.

These examples demonstrate the versatility of complex text-based GIAs, showcasing their ability to process and generate text in ways that are transformative for various industries and domains. As AI technology continues to advance, the scope and impact of these agents are likely to expand even further, offering more sophisticated and nuanced applications.

Spotlight on Complex Text-Based GIAs for Strategy

Complex text-based GIAs for strategy represent a cutting-edge application of AI, particularly in areas that require sophisticated decision-making and strategic planning. These advanced systems leverage their generative and analytical capabilities to assist in formulating, evaluating, and optimizing strategies in various domains such as business, military, finance, and game theory.

Capabilities of Complex Text-Based GIAs in Strategy

Strategic Analysis and Insight Generation: These GIAs can analyze vast amounts of textual data, including market reports, financial documents, and news articles, to identify trends, opportunities, and threats. They can then generate insights and strategies that are coherent, contextually relevant, and actionable.

Scenario Planning and Simulation: They can simulate multiple strategic scenarios by generating text-based narratives and outcomes based on different variables. This helps in understanding potential future developments and preparing for various contingencies.

Decision Support: By processing and synthesizing information, these GIAs can provide recommendations and decision support to strategists and decision-makers. They can evaluate the potential outcomes of different strategic choices and suggest optimal paths.

Dynamic Learning and Adaptation: As they interact with users and consume new data, these GIAs can learn and adapt, improving their strategic recommendations over time. They can incorporate feedback and new information to refine their understanding of strategic landscapes.

Natural Language Interaction: Their ability to interact in natural language makes them user-friendly and accessible to a wide range of professionals, not just those with technical backgrounds. This allows for broader implementation and collaboration across various departments and levels within an organization.

Real-Time Data Processing and Response: They can process real-time data feeds and provide immediate strategic insights, which is crucial for fast-paced environments like financial trading or crisis management.

Applications of Complex Text-Based GIAs in Strategy

Business Strategy and Competitive Analysis: In the corporate sector, these GIAs can analyze market trends, competitor strategies, and customer feedback to help businesses develop effective strategies.

Military and Defense Strategy: They can be used to simulate conflict scenarios, analyze military intelligence, and assist in formulating defense strategies.

Financial Market Analysis: In finance, these agents can analyze market data, economic reports, and news to provide investment strategies and risk assessments.

Policy Making and Governance: Governments and NGOs can use these systems to analyze policy impacts, public opinion, and socioeconomic data to formulate and adjust policies.

Game Theory and Decision Science: They can be used to model game-theoretic scenarios and decision-making processes, providing insights into optimal strategies in competitive environments.

Challenges and Considerations

Accuracy and Reliability: The effectiveness of these GIAs depends on the quality of their training data and algorithms. Inaccurate or biased data can lead to flawed strategies.

Ethical and Privacy Concerns: The use of sensitive data for strategic planning raises ethical and privacy issues. It's crucial to ensure data is used responsibly and in compliance with regulations.

Human Oversight: Despite their advanced capabilities, the need for human oversight remains essential. Strategic decisions often have significant consequences, and relying solely on AI without human judgment and contextual understanding can be risky.

Complex Text-Based Generative Intelligent Agents are transforming the landscape of strategic planning and decision-making. Their ability to process vast amounts of data, generate strategic insights, and adapt to new information makes them invaluable tools in various fields. However, the successful implementation of these systems requires careful consideration of their limitations, ethical implications, and the need for human oversight. As these technologies continue to evolve, they hold the potential to significantly enhance the sophistication and effectiveness of strategic planning processes.

Fine-Tuning Complex Text-Based GIAs

Fine-tuning complex text-based GIAs is a critical process to enhance their performance, accuracy, and relevance to specific tasks or domains. This process involves adjusting and customizing pretrained models to better suit the unique requirements of a particular application. The following is an overview of how fine-tuning is typically carried out.

Understanding Pretrained Models

Base Models: GIAs often start with a base model trained on a large, generalized dataset. This model has a broad understanding of language but lacks specialization.

Examples: Models like OpenAI's GPT-4 or Claude 2 by Anthropic are commonly used as starting points due to their extensive training on diverse text corpora.

Fine-Tuning Process

Selecting a Target Dataset: The dataset for fine-tuning should be closely related to the specific task or domain, for example, legal documents for a legal GIA or customer interactions for a customer-service GIA.

Data Preprocessing: The selected data needs to be cleaned and structured appropriately. This might include removing irrelevant data, correcting errors, and ensuring data consistency.

Transfer Learning: This involves adapting the pretrained model to the new dataset. The model is exposed to the new data, allowing it to learn the nuances, terminology, and patterns specific to the target domain.

Hyperparameter Adjustment: Parameters like learning rate, batch size, and the number of training epochs are adjusted to optimize the learning process.

Model Training: The GIA is trained on the new dataset until it achieves satisfactory performance. This may involve several iterations to find the right balance between retaining general language skills and acquiring specialized knowledge.

Evaluation and Validation: The fine-tuned model is tested using a separate validation dataset to ensure it performs well on unseen data. Metrics such as accuracy, recall, precision, and F1 score are typically used for evaluation.

Challenges and Considerations

Overfitting: There's a risk that the model becomes too specialized to the training data and fails to generalize well. Regularization techniques and proper validation can help mitigate this.

Data Quality and Bias: The quality and representativeness of the training data are crucial. Biased or poor-quality data can lead to a biased or underperforming GIA.

Continuous Learning: Post-deployment, the GIA may need continuous updates and retraining to maintain its relevance, especially in fast-changing fields.

Ethical Considerations: Care must be taken to ensure that the fine-tuning process does not introduce or amplify unethical biases, especially when dealing with sensitive applications.

Applications

Fine-tuning allows GIAs to be effectively applied in diverse fields:

- **Healthcare**: For accurate medical diagnoses or patient interaction.

- **Finance**: For market analysis or personalized financial advice.

- **Legal**: For legal research or contract analysis.

- **Customer Support**: For industry-specific customer interactions.

Fine-tuning is a crucial step in deploying complex text-based GIAs, ensuring that they are not only proficient in language but also experts in their specific domain of application. This process requires careful consideration of the data used, the model's capabilities, and the specific needs of the application to achieve the best results. As AI technology continues to advance, the fine-tuning process will become even more sophisticated, leading to more accurate and efficient GIAs across various sectors.

Future of Complex Text-Based GIAs

The future of complex Text-Based Generative Intelligent Agents (GIAs) is poised to be a fascinating and transformative journey, with these advanced systems expected to integrate more deeply into various aspects of human life and industry. The following is a glimpse into what the future might hold for these AI entities.

Enhanced Interactivity and Personalization

Conversational AI: Future GIAs will likely exhibit more nuanced and contextually rich conversational abilities, making interactions with AI almost indistinguishable from human interactions.

Deep Personalization: Advanced algorithms will enable GIAs to tailor their responses and content generation to individual user preferences and behaviors, enhancing the user experience in applications like virtual assistance, customer service, and personalized learning.

Broader Integration Across Industries

Healthcare: GIAs could assist in diagnostic processes, patient care, and even therapeutic settings, offering personalized advice and support based on patient history and current medical data.

Education: AI tutors could provide personalized education, adapting to each student's learning style and pace, and generating custom learning materials.

Legal and Compliance: AI-driven legal research and analysis will become more sophisticated, offering more accurate predictions of case outcomes and automating complex documentation.

Creative and Intellectual Contributions

Content Creation: We will see GIAs contributing creatively, writing books, composing music, or creating art, with a level of sophistication that may challenge our understanding of creativity.

Research and Development: GIAs could autonomously conduct research, propose hypotheses, and even author scientific papers, accelerating innovation across scientific fields.

Ethical AI and Bias Mitigation

Ethical Frameworks: As GIAs become more integrated into daily life, the development of robust ethical frameworks to guide their decision-making processes will be crucial.

Bias Detection and Correction: Advanced techniques will likely be developed to identify and mitigate biases in AI-generated content and interactions.

Advanced Learning Capabilities

Continuous Learning: Future GIAs will continually learn from interactions and external data sources, constantly updating and refining their knowledge base and models.

Cross-Domain Knowledge: GIAs may evolve to possess cross-domain knowledge, allowing them to provide insights and generate content that spans multiple disciplines.

User Interface Evolution

Beyond Text: The interaction with GIAs will accelerate beyond text, easily incorporating voice, visual elements, and even VR or AR interfaces.

Seamless Integration: GIAs could become seamlessly integrated into everyday objects and environments, offering assistance and generating content on the go.

Societal and Regulatory Adaptations

Regulatory Frameworks: New laws and regulations will likely be established to govern the use and output of GIAs, particularly in areas like media, education, and legal advice.

Impact on Employment: As GIAs take on more complex tasks, there will be shifts in job markets and employment, necessitating new strategies for workforce adaptation and skills development.

The future of complex text-based GIAs holds immense potential and is likely to impact almost every aspect of human endeavor. As these technologies advance, they will challenge our concepts of creativity, productivity, and even the nature of human–AI interaction.

Summary

Text-Based Generative Intelligent Agents (GIAs) have marked their territory far beyond the realms traditionally occupied by chatbots and virtual assistants. These advanced AI constructs, powered by Generative AI, have not only redefined the interaction between humans and machines but have also opened a gateway to an era of unbounded creativity and innovation. As we move forward, these agents are set to become an integral part of our digital infrastructure, offering personalized, interactive experiences and generating content with a degree of sophistication that blurs the lines between human and machine intelligence.

In the sphere of language and textual content, GIAs have demonstrated their potential to go beyond mere conversation, venturing into the creation of written works that resonate with human emotions and intellect. Their ability to grasp context, idioms, and subtleties of language enables them to produce text that is not only coherent but also rich in the complexities that characterize human writing. From drafting legal documents to composing poetic verses, GIAs have shown that they can adapt their output to fit a vast spectrum of user needs and preferences.

The implications of such technology are profound. In journalism, business intelligence, and education, GIAs are transforming the way content is created and consumed. They are becoming indispensable tools for companies looking to scale their content strategies without compromising on quality or losing the personal touch that engages audiences. In customer service, GIAs offer a level of interaction and problem-solving capability that elevates the customer experience, making it more efficient and responsive.

However, with great power comes great responsibility. As GIAs become more prevalent, ethical concerns and the challenge of maintaining authenticity in AI-generated content must be addressed. The dependency on data quality and the potential perpetuation of biases present an ongoing challenge that requires vigilance and continuous improvement of the underlying models and training methods.

The distinction between GIAs and their less sophisticated counterparts, traditional chatbots, and virtual assistants, is stark. While the latter are constrained by the limits of their programming, GIAs thrive on their ability to learn, adapt, and generate. As they become more advanced, these agents are expected to take on more complex roles, aiding in decision-making processes and strategy formulation across various sectors.

The versatility of GIAs is already evident, with applications ranging from market analysis to creative writing, and their future is even more promising. As these systems evolve, we can anticipate a world where GIAs collaborate with humans in nearly every creative and intellectual endeavor, augmenting our capabilities and expanding the horizons of what can be achieved.

In conclusion, Text-Based Generative Intelligent Agents stand on the threshold of a new digital dawn. They promise to enrich our interactions with technology, push the boundaries of creativity, and redefine the limits of automated systems.

CHAPTER 13

Applications and Real-World Case Studies

Generative AI stands as a transformative force in modern business, for example, infusing Decision Support Systems (DSS) and Autonomous Systems with advanced capabilities for risk assessment and strategic decision-making. In financial sectors, it revolutionizes risk management by generating detailed synthetic data and simulating complex economic scenarios, enabling institutions to forecast market trends and navigate economic fluctuations with enhanced precision. This technology has become critical for stress testing, portfolio management, and ensuring regulatory compliance.

Extending its reach beyond finance, Generative AI serves as the backbone of operational risk management by modeling potential disruptions in supply chains and IT systems, thus empowering businesses to proactively mitigate vulnerabilities. It plays a strategic role in planning, allowing companies to test the waters of market entry, product launches, and operational overhauls before fully committing to new business strategies.

Generative AI's predictive prowess also fortifies cybersecurity efforts, simulating digital threats to help companies bolster their defenses and mount timely responses to cyberattacks. In the realm of disaster preparedness, it projects the impact of catastrophic events, underpinning robust response strategies and business continuity planning.

Moreover, in DSS, Generative AI aids decision-makers by simulating outcomes to complex problems, providing a range of potential solutions and outcomes. It's instrumental in crafting user-tailored interfaces that present data-driven insights for informed decisions. In Autonomous Systems, from self-navigating warehouses to self-driving vehicles, Generative AI optimizes routes, predicts maintenance needs, and ensures efficient operations.

Generative AI is rapidly becoming central to corporate strategy, providing businesses with the tools to confidently face an uncertain future. Its integration into DSS and Autonomous Systems showcases its breadth, offering a competitive advantage and driving innovation in a rapidly evolving business ecosystem.

© Irena Cronin 2024
I. Cronin, *Understanding Generative AI Business Applications*, https://doi.org/10.1007/979-8-8688-0282-9_13

Business Simulations for Risk Assessment

Generative AI is increasingly becoming a valuable tool for businesses in conducting simulations for risk assessment. Its capability to analyze data and predict outcomes based on that data is particularly useful in creating simulations that can forecast a variety of business risks.

Financial Risk Assessment

Generative AI's application in financial risk assessment is a significant leap forward in the way financial institutions manage and mitigate risk. By using these AI models, companies can gain insights that were previously inaccessible due to the limitations of traditional statistical methods. The following is an in-depth look at how Generative AI is transforming financial risk assessment:

Creating Synthetic Financial Data:

Generative AI, particularly models like Generative Adversarial Networks (GANs), can create synthetic financial data that is statistically similar to real historical data. This synthetic data can fill gaps in incomplete datasets, allowing for more comprehensive stress testing and scenario analysis.

Scenario Analysis and Stress Testing:

Financial institutions are required to conduct stress tests to evaluate how their portfolios would perform under adverse conditions. Generative AI can simulate a wide array of economic conditions—from market crashes to slow economic downturns—to help institutions prepare for and potentially mitigate the effects of financial downturns.

Predictive Analysis for Market Trends:

By training on vast amounts of historical data, Generative AI models can identify patterns and correlations that may not be apparent to human analysts. These models can extrapolate this data to predict future market trends, helping companies to anticipate market movements and adjust their strategies accordingly.

Risk Management for Investment Portfolios:

Generative AI can simulate the behavior of different assets under various market conditions to identify potential risks in investment portfolios. This can inform better asset allocation, as well as the development of hedging strategies to protect against potential losses.

Tail Risk Estimation:

Generative AI can be particularly useful in estimating tail risk—the risk of rare but extreme market events. By generating data for events that have little historical precedent, AI models can help financial analysts better understand and prepare for these outliers.

Regulatory Compliance:

Regulators are increasingly expecting financial institutions to have sophisticated risk assessment capabilities. Generative AI can help ensure compliance with these regulations by providing detailed analyses of potential risks and demonstrating that institutions are prepared for a range of outcomes.

Optimization of Trading Strategies:

Traders can use Generative AI to simulate trading strategies under different market conditions to find the most robust approaches. This can lead to the development of strategies that are more resilient to market volatility.

Credit Risk Modeling:

In the lending space, Generative AI can enhance credit risk models by simulating the impact of economic scenarios on borrowers' ability to repay loans. This can lead to more accurate credit scoring and better-informed lending decisions.

Generative AI provides financial institutions with a powerful tool to forecast and prepare for financial risks. These AI models offer a level of depth and flexibility in analysis that can greatly enhance decision-making processes in financial risk assessment, leading to more resilient financial operations and strategies. As these technologies continue to advance, they will likely become an integral part of financial planning and risk management.

Operational Risk Management

Generative AI can play a crucial role in operational risk management by simulating a wide array of internal processes and external events that could impact a company's operations.

Supply Chain Optimization:

Generative AI can create simulations that model the entire supply chain process, allowing companies to anticipate the impact of various disruptions such as supplier failures, transportation delays, or changes in demand. By analyzing these simulations, companies can identify critical chokepoints in their supply chain and develop contingency plans to ensure continuity.

Predicting Machinery Breakdowns:

With predictive maintenance, Generative AI can forecast when machinery is likely to fail or require maintenance. This is achieved by analyzing operational data and identifying patterns that precede equipment failures. By addressing these issues proactively, companies can reduce downtime and associated costs.

System Failures and IT Risk:

In the realm of IT, Generative AI can simulate network outages, cyberattacks, and other system failures to help IT departments strengthen their systems against such incidents. These simulations can inform better system design, the implementation of robust cybersecurity measures, and disaster recovery planning.

Process Improvement:

Generative AI can also simulate various operational processes to identify inefficiencies and areas for improvement. By modeling different scenarios, companies can streamline operations, reduce waste, and improve overall efficiency.

Risk Quantification:

Operational risks can be difficult to quantify. Generative AI can help by simulating the financial impact of operational failures, providing companies with a clearer understanding of the potential costs associated with different risk scenarios.

Human Factors and Safety:

Generative AI can be used to model scenarios involving human factors, such as errors or accidents, to improve workplace safety. By understanding how human behavior can lead to operational risks, companies can develop better training programs and safety protocols.

Regulatory Compliance:

Companies can use Generative AI to ensure compliance with regulatory requirements. By simulating the impact of regulatory changes on operations, companies can adapt their processes to remain compliant while minimizing disruption to their business.

Scenario Planning and Decision-Making:

Generative AI allows companies to engage in comprehensive scenario planning. By simulating a wide range of possible futures, companies can better prepare for uncertainty and make more informed strategic decisions.

Generative AI's ability to simulate and predict the outcomes of complex operational scenarios is a game-changer for operational risk management. It enables businesses to proactively identify vulnerabilities, optimize processes, and implement strategies to mitigate risks, thereby safeguarding against potential disruptions and losses. As businesses continue to face a rapidly changing operational environment, Generative AI becomes an invaluable tool for navigating these complexities and maintaining a competitive edge.

Strategic Planning

Generative AI holds significant potential for enhancing strategic planning across various business domains. By simulating different business strategies, Generative AI allows companies to navigate potential futures and devise robust, data-driven plans. The following is a closer look at the applications and benefits of Generative AI in strategic planning:

Market Entry Strategies:

When a company considers entering a new market, there are numerous variables to consider, such as customer behavior, competition, regulatory environment, and cultural nuances. Generative AI can simulate market entry scenarios to help companies understand potential barriers and the competitive landscape. By assessing the risk and reward profiles of different strategies, businesses can determine which markets are worth entering and the best approach to do so.

New Product Launches:

Launching a new product involves considerable risk due to uncertainties about market demand, pricing strategies, and competitive response. Generative AI can simulate the market's reaction to a new product, including acceptance and sales projections, helping companies refine their product offerings and marketing strategies before launch.

Business Operations Changes:

Changes in business operations, such as shifts in the supply chain, manufacturing processes, or distribution methods, can have significant implications for a company's bottom line. Generative AI can model the outcomes of operational changes, providing insights into the potential cost savings, efficiency gains, or potential new bottlenecks.

Financial Forecasting:

Strategic planning often requires accurate financial forecasting. Generative AI can simulate different financial scenarios based on various strategic decisions, such as changes in investment, cost-cutting measures, or capital allocation. This helps businesses to forecast revenues, expenses, and profitability under different conditions.

Scenario Analysis:

Generative AI excels at creating numerous "what-if" scenarios, such as changes in the economic environment, shifts in consumer preferences, or new technological disruptions. Companies can use these simulations to test the resilience of their strategies against possible future events.

Mergers and Acquisitions:

For companies looking to grow through mergers and acquisitions, Generative AI can simulate the integration process and predict the synergies, costs, and potential risks involved. This helps in evaluating the feasibility and long-term impact of such strategic moves.

Crisis Management:

Generative AI can also be utilized for crisis management planning. By simulating crisis scenarios, companies can plan their responses and develop strategies to mitigate risks associated with reputational damage, operational disruption, or financial losses.

Long-term Planning:

For long-term strategic planning, Generative AI can simulate the broader impacts of technological trends, demographic shifts, and global economic changes on a company's business model. This forward-looking approach helps companies to align their long-term goals with the evolving market and technological landscape.

Generative AI serves as a strategic asset for companies, offering a powerful tool for simulating and evaluating the multitude of factors that affect strategic decisions. By leveraging these AI-driven insights, companies can craft more informed, adaptive, and resilient strategic plans, positioning themselves effectively for the future.

Cybersecurity Threat Analysis

Generative AI is becoming an essential tool in the realm of cybersecurity threat analysis. As cyber threats grow more sophisticated, traditional security measures often struggle to keep up. How Generative AI is being used to enhance cybersecurity defenses is shown as follows:

Generating Simulated Attacks:

Generative AI can create realistic cyberattack simulations that mimic the tactics, techniques, and procedures of actual hackers. By doing so, it helps organizations understand how an attacker might exploit vulnerabilities in their systems. This proactive approach allows companies to identify and patch potential security loopholes before they can be exploited.

Training and Preparedness:

By using Generative AI to simulate attacks, companies can better train their cybersecurity personnel. Security teams can engage with these simulations to practice their response to a variety of attack scenarios, thereby improving their readiness and response strategies in case of a real incident.

Threat Modeling:

Generative AI can help in threat modeling by predicting potential attack vectors. It can take into account the ever-changing cybersecurity landscape and anticipate future threats based on emerging patterns. This predictive capability allows organizations to prepare for and potentially prevent attacks that have not yet been observed in the wild.

Enhancing Detection Systems:

Generative AI can improve intrusion detection systems by generating new patterns of malicious activity that haven't been seen before. This helps in updating the detection algorithms to identify and respond to novel attack methods.

Security Testing of AI Systems:

As AI systems are increasingly integrated into various aspects of business operations, ensuring their security is paramount. Generative AI can be used to test the robustness of other AI systems against adversarial attacks, ensuring these systems can resist manipulation and function correctly even when under attack.

Phishing and Social Engineering Simulations:

Phishing attacks and other forms of social engineering are constantly evolving. Generative AI can simulate these attacks to test the effectiveness of organizational training and employee awareness programs, as well as to improve filtering algorithms to better detect and block such attempts.

Analyzing Ransomware Tactics:

Ransomware remains one of the most significant threats to businesses. Generative AI can simulate ransomware attacks to help organizations better understand how their systems could be held hostage and develop more effective backup and recovery strategies to mitigate the impact of such attacks.

Tailoring Defenses to Specific Threats:

Every organization has a unique digital infrastructure, which means that a one-size-fits-all approach to cybersecurity is often insufficient. Generative AI can tailor simulated attacks to the specific configurations and vulnerabilities of an organization's infrastructure, leading to customized defense mechanisms that are far more effective.

Generative AI serves as a sophisticated tool for simulating and analyzing cybersecurity threats, enabling organizations to fortify their defenses against an ever-evolving array of cyber threats. By incorporating Generative AI into their cybersecurity strategies, organizations can enhance their ability to predict, prevent, and respond to cyberattacks, safeguarding their data and systems in a proactive manner.

Disaster Preparedness

Generative AI is significantly changing how businesses prepare for disasters by providing sophisticated simulation capabilities for disaster preparedness and risk assessment. The following are the ways in which Generative AI contributes to this critical aspect of business continuity planning:

Disaster Scenario Modeling:

Generative AI can model a wide range of disaster scenarios, from natural calamities like earthquakes and floods to man-made crises such as industrial accidents or acts of terrorism. These models can predict the sequence of events that might occur during such disasters, helping businesses to understand potential impacts on their operations. Generative AI could be used to create synthetic data which could be used for scenario creation to greatly aid in prediction (e.g., `www.gao.gov/products/gao-24-106213`).

Business Continuity Planning:

By simulating the effects of various disaster scenarios, Generative AI aids in the creation and refinement of business continuity plans. It allows businesses to test different response strategies and identify the most effective ways to maintain critical operations during and after a disaster.

Resource Allocation:

Generative AI can help businesses optimize their allocation of resources for disaster response. By simulating the demand for resources in different scenarios, companies can better plan for the supplies, personnel, and equipment needed to ensure resilience.

Supply Chain Disruptions:

Generative AI can anticipate the effects of disasters on the supply chain, identifying which suppliers, transportation routes, or logistics hubs are most at risk. This enables companies to develop alternative supply chain strategies that can be activated in the event of a disruption.

Infrastructure Analysis:

Companies can use Generative AI to analyze the resilience of their physical infrastructure against disasters. This can range from the stability of buildings in an earthquake to the risks posed to data centers by flooding.

Emergency Response Training:

Generative AI can create simulations for training purposes, allowing emergency response teams to practice their skills in a virtual environment that mimics real-life disaster conditions. This hands-on experience can be invaluable in preparing teams for actual emergencies.

Insurance and Financial Planning:

With Generative AI, businesses can simulate the financial impact of disasters, which can inform insurance coverage decisions and financial preparations. This ensures that companies have adequate coverage and financial reserves to recover from potential losses.

Communication Strategies:

Effective communication is crucial during a disaster. Generative AI can simulate different communication challenges and help businesses develop strategies to ensure that critical information reaches employees, customers, and stakeholders promptly and efficiently.

Recovery Time Estimation:

Generative AI can estimate how long it will take for various aspects of the business to recover after a disaster. This helps in setting realistic expectations and planning for the post-disaster recovery phase.

Generative AI is a powerful tool that equips businesses with the ability to simulate and prepare for a range of disaster scenarios. By leveraging these advanced simulations, companies can develop robust preparedness strategies, mitigate risks, and ensure that they are able to maintain continuity and recover swiftly in the face of adverse events. This preparedness is crucial not only for the survival of the business but also for the safety and well-being of its employees and the communities it serves.

Customer Behavior Modeling

Customer behavior modeling using AI involves using data-driven techniques to predict how customers are likely to behave under various circumstances. Generative AI, with its ability to create realistic simulations, can be particularly effective in this area. Several ways in which Generative AI enhances customer behavior modeling are as follows:

Predicting Purchase Behavior:

Generative AI can simulate customer reactions to different products and services, which helps businesses understand what drives purchase decisions. These simulations can take into account factors like price changes, new features, competitor actions, and market conditions.

Personalization of Marketing Strategies:

By modeling customer behavior, Generative AI enables businesses to tailor their marketing strategies to individual preferences. This can include personalized recommendations, targeted advertising, and customized promotions, all designed to increase customer engagement and conversion rates.

Demand Forecasting:

AI models are adept at predicting future demand for products by analyzing past buying patterns, seasonal trends, and other relevant data. This forecasting is critical for effective inventory management, ensuring that businesses stock the right amount of product to meet customer demand without overstocking.

Sentiment Analysis:

Generative AI can simulate customer sentiment and emotional responses to brands and marketing campaigns. This insight allows companies to adjust their messaging and branding to better resonate with their target audience.

Price Elasticity Modeling:

Understanding how sensitive customers are to price changes is crucial for setting pricing strategies. Generative AI can model scenarios of customer response to various pricing strategies, helping businesses find the optimal balance between profitability and customer satisfaction.

Churn Prediction:

AI can simulate scenarios to predict customer churn, which is when a customer stops doing business with a company. By identifying patterns that indicate a customer is at risk of leaving, businesses can take proactive measures to retain them.

New Product Development:

In product development, Generative AI can help simulate how customers might perceive a new product, including which features they might find most appealing and what potential pain points could exist. This can guide the product design process to better align with customer needs and preferences.

Understanding Customer Journeys:

Generative AI can simulate the various paths customers take from discovering a product to making a purchase. Understanding these customer journeys can help businesses optimize the sales funnel and remove obstacles that might deter customers.

A/B Testing Simulations:

Instead of conducting live A/B tests, which can be costly and time-consuming, Generative AI can simulate the outcomes of A/B tests for different website layouts, marketing copy, or product features, allowing businesses to test more hypotheses in less time.

Generative AI serves as a sophisticated tool for customer behavior modeling, enabling businesses to gain a deeper understanding of their customers and predict their behaviors with greater accuracy. This predictive power can lead to more informed business decisions, from product development to marketing and beyond, ensuring that companies remain competitive in a market that is increasingly driven by consumer insights and data.

Decision Support Systems: Tools and Technologies

In the context of Decision Support Systems (DSS), Generative AI can be leveraged in numerous ways to enhance decision-making processes, including risk assessment which we just covered.

Predictive Analytics

Generative AI models, such as neural networks and machine learning algorithms, can analyze historical data to predict future outcomes. For instance, they can forecast sales trends, customer behaviors, or market movements, providing decision-makers with actionable insights.

Scenario Planning

These AI tools can simulate a range of possible future scenarios based on varying inputs and conditions. This helps organizations evaluate the potential impacts of different strategies or decisions, taking into account a variety of external factors like economic shifts or competitive actions.

Advanced Simulations

Generative AI models can simulate highly realistic data for training simulations in fields like medicine or aviation, improving the quality of training without the risks associated with real-world training.

Data Augmentation

Generative AI can create additional synthetic data when real data is scarce. This is particularly useful in situations where data collection is difficult or expensive, such as rare event prediction or dealing with sensitive information.

Optimization

Generative AI can assist in optimization problems, finding the best solution from a set of possibilities. It can optimize routes for logistics, resource allocation for projects, or even portfolio management for finance.

Natural Language Processing (NLP)

Generative AI with NLP can extract insights from unstructured data sources like social media, customer reviews, or news articles, which can inform decisions regarding market sentiment, brand reputation, or emerging trends.

Content Generation

In marketing and communications, Generative AI can create drafts for reports, summaries, or even marketing content, which can then be refined by human experts. This helps in speeding up the content creation process and providing a starting point for further analysis.

Decision Automation

In some cases, Generative AI can automate routine decision-making processes by generating decisions based on predefined criteria. This allows human decision-makers to focus on more complex, high-level strategic decisions.

User Interface and Interaction

Generative AI can enhance user interfaces in DSS by generating personalized dashboards and reports that align with the specific preferences or needs of the decision-maker.

By incorporating these capabilities into DSS, Generative AI can significantly improve the quality and speed of decision-making across all levels of an organization. It enables businesses to harness the power of their data, anticipate future trends, and make decisions that are informed by a deep understanding of potential outcomes.

Autonomous Systems: From Warehouses to Vehicles

Generative AI has wide-ranging applications for autonomous systems, encompassing everything from optimizing warehouse operations to enhancing the functionality of autonomous vehicles.

Enhancing Autonomous Navigation

In autonomous vehicles, Generative AI can be used to simulate countless driving scenarios, which the vehicle's AI system may not have encountered in the real world. This includes challenging weather conditions, unexpected obstacles, or unique traffic patterns. By training on these simulations, autonomous driving systems can improve their decision-making algorithms and enhance safety.

Predictive Maintenance

For both warehouses and vehicles, Generative AI can predict when a component is likely to fail by analyzing operational data. This predictive maintenance helps to minimize downtime and extend the lifespan of equipment.

Route and Layout Optimization

In warehouses, Generative AI can optimize the layout for storage and the paths that autonomous robots take to pick and place items. For autonomous vehicles, similar algorithms can find the most efficient routes in real time, considering traffic conditions and other variables.

Load Balancing

Generative AI can simulate different load distributions within a warehouse or a vehicle to find the optimal balance that maximizes efficiency and safety. This can lead to improved handling and fuel efficiency in vehicles, as well as better utilization of space in warehouses.

Traffic Flow Optimization

For autonomous vehicles, particularly in urban environments, Generative AI can simulate various traffic conditions to optimize flow and reduce congestion. These simulations can inform the development of traffic management systems that facilitate smoother rides and faster travel times.

Human Interaction Training

Generative AI can simulate interactions between autonomous systems and humans, whether it's pedestrians in the case of autonomous vehicles or workers in the case of warehouse robots. This helps the AI learn to anticipate and react to human behavior safely and effectively.

Energy Efficiency

Generative AI can simulate and optimize energy usage patterns to make autonomous systems more efficient. In autonomous electric vehicles, for instance, this can lead to algorithms that extend the range of the vehicle by optimizing driving styles and energy consumption.

Anomaly Detection

Generative AI can learn what "normal" operation looks like for an autonomous system and then generate scenarios of various anomalies. This aids in the early detection of issues that could lead to system failures or unsafe conditions.

Decision-Making Under Uncertainty

Autonomous systems often have to make decisions with incomplete information. Generative AI can simulate uncertain environments and outcomes, helping autonomous systems improve their decision-making algorithms for these situations.

Adaptive Learning

As environments and tasks change, autonomous systems need to adapt. Generative AI allows these systems to learn from simulated environments that are constantly changing, ensuring that the systems can adapt to new tasks or conditions without direct human intervention.

Summary

In summary, Generative AI acts as a force multiplier for the capabilities of autonomous systems, from improving operational efficiency to ensuring safety and reliability. By leveraging the power of simulation and predictive analytics, Generative AI empowers these systems to perform more effectively and adapt to the complex and dynamic environments in which they operate.

In conclusion, the real-world applications are numerous and growing, with case studies demonstrating Generative AI's role. Generative AI has emerged as a linchpin in the machinery of modern business, powering Decision Support Systems and Autonomous Systems with unprecedented foresight and adaptability. It has revolutionized financial risk assessment, enabling deeper insights into market trends and regulatory compliance through sophisticated simulations. Generative AI extends its utility into operational risk management, preempting supply chain disruptions, and fortifying cybersecurity defenses against evolving threats. Strategic planning now benefits from AI's ability to forecast the ramifications of business decisions, ensuring companies navigate market entries and product launches with confidence.

As the corporate world grapples with uncertainties, the strategic integration of Generative AI across various sectors is not just an advantage but a necessity, solidifying its status as a transformative force in the landscape of global business.

CHAPTER 14

Summarizing Key Insights

In the unfolding narrative of technological evolution, Generative AI emerges as a pivotal chapter, marking the confluence of computational precision and human-like creativity. This transformative technology is redefining the boundaries of artificial intelligence by enabling the autonomous generation of new content that spans text, images, audio, and video. The leap from traditional AI, with its focus on analysis and interpretation, to one that actively creates and designs, is akin to a paradigm shift, mirroring the complex processes of human ingenuity. Through the intricate interplay of advanced algorithms and extensive datasets, Generative AI is poised to unlock groundbreaking potential across various industries, offering novel solutions and reshaping the way we perceive machine capabilities.

At the heart of this innovative surge are the technical underpinnings of machine learning, deep learning, and, notably, neural networks, with transformer models standing at the forefront of this revolution, especially in handling natural language processing tasks with remarkable finesse. These developments not only fuel efficiency and innovation in business practices but also revolutionize human–computer interactions through chatbots and virtual assistants that provide unprecedented personalization. Here, we review the multifaceted dimensions of Generative AI, examining its profound impact on content creation, decision-making processes, and its broader applications that extend beyond mere functionality to an art form that blurs the line between the digital and the tangible.

Highlights from Each Section

Generative AI represents a transformative class of AI technology designed to autonomously generate new content, including but not limited to text, images, audio, and video. This capability marks a significant shift from traditional AI's focus on understanding or interpreting content to actively creating it, resembling the creative

© Irena Cronin 2024
I. Cronin, *Understanding Generative AI Business Applications*, https://doi.org/10.1007/979-8-8688-0282-9_14

processes associated with human intelligence. Generative AI leverages complex algorithms and vast datasets to produce outputs that can mimic human-like creativity, offering groundbreaking possibilities across various sectors.

Core Technical Concepts

At the heart of Generative AI lie advanced technical frameworks, primarily based on ML, deep learning, and particularly, neural networks. Among these, transformer models stand out for their effectiveness in handling NLP tasks, enabling machines to understand, interpret, and generate human language with remarkable proficiency. These models rely on architectures that can process sequences of data, such as sentences, making them ideal for a range of applications from translation services to generating novel textual content.

The Business Case for Generative AI

The advent of Generative AI presents compelling opportunities for businesses across industries. By automating creative processes, companies can achieve higher efficiency, reduce costs, and create personalized experiences for their customers. Generative AI applications in content creation, customer service, and product design not only streamline operations but also open new avenues for innovation, enabling businesses to stay competitive in a rapidly evolving digital landscape.

The World of Text-Based Generative AI

In the domain of text, Generative AI technologies are revolutionizing the way we interact with machines. Chatbots and virtual assistants powered by sophisticated NLP models offer human-like interactions, providing customer support, personalized recommendations, and automated communication services. Beyond simple interactions, these technologies extend to document automation, sentiment analysis, and AI-driven content creation, significantly enhancing productivity and engagement in content-heavy sectors.

Unpacking Transformer-Based NLP

Transformer-based NLP represents a leap forward in understanding and generating human language. Unlike previous models that processed text linearly, transformers analyze entire sequences of words in parallel, capturing the nuances of language more effectively. This breakthrough has enabled more coherent and contextually relevant text generation, laying the foundation for advanced chatbot technologies and other text-based applications.

Exploring Chatbot Technologies

Chatbot technologies have evolved from simple rule-based systems to complex AI-driven interfaces capable of conducting sophisticated conversations. By leveraging transformer-based NLP, chatbots can understand queries, infer context, and provide responses that closely mimic human interaction, transforming customer service and engagement across digital platforms.

Advanced Applications of Text-Based AI

Text-based AI is not limited to communication; it extends to document automation, where AI systems can draft, review, and manage documents, significantly reducing manual effort. Sentiment analysis, another application, allows businesses to analyze customer feedback and social media discourse, gaining insights into public sentiment. AI-driven content creation tools are redefining marketing and media, generating articles, reports, and content at scale, as well as aiding in code generation which many software engineers are taking advantage of.

Senses-Based Generative AI Demystified

Beyond text, Generative AI encompasses sense-based applications, including visual and auditory technologies. Innovations in visual algorithms, such as NeRFs and 3D Gaussian Splatting, are creating lifelike 3D models and environments from 2D images, enhancing VR, AR, and computer vision applications. Text-to-image and text-to-video synthesis further expand the creative possibilities, enabling the generation of visual content directly from textual descriptions.

In-depth Look at Visual Algorithms

Visual algorithms are at the forefront of transforming how businesses leverage computer vision. Techniques like NeRFs create highly realistic 3D environments, while text-to-image synthesis generates visual content from textual prompts, offering novel ways to engage users and visualize data. These advancements are increasingly applied in business settings, from product visualization to marketing campaigns, offering unprecedented opportunities for innovation.

The Auditory and Multisensory Experience

Generative AI also extends to the auditory domain, synthesizing music, sound effects, and voiceovers, thereby enriching multimedia content and interactive experiences. In combination with AR and VR, these auditory applications provide immersive environments for training, simulation, and entertainment, bridging the gap between digital and physical experiences.

Rationale-Based Generative AI

Moving beyond content generation, rationale-based Generative AI focuses on simulating complex decision-making processes. This aspect of Generative AI finds application in scenarios ranging from financial analysis to manufacturing planning, where AI systems can model various outcomes based on extensive datasets, aiding in strategic decision-making and risk management.

Data and Algorithms: The Foundation

The capabilities of Generative AI are underpinned by sophisticated data analysis and machine learning algorithms. These technologies enable the processing and interpretation of large volumes of data, facilitating the training of AI models that can predict trends, automate decision-making, and drive innovation across sectors.

Applications and Real-World Case Studies

Generative AI's real-world applications are vast and varied, encompassing business simulations for risk assessment, autonomous systems for logistics and transportation, and decision support systems for strategic planning. These applications demonstrate the potential of Generative AI to revolutionize industries, offering new efficiencies, enhancing decision-making, and creating innovative products and services.

In essence, Generative AI represents a frontier in artificial intelligence, bridging the gap between machine efficiency and human creativity. Its applications across text, senses, and decision-making showcase the technology's versatility and potential to redefine industries. As businesses and developers continue to explore its capabilities, Generative AI is set to drive the next wave of digital transformation, offering profound implications for how we work, create, and interact with the world around us.

Business Benefits and Drawbacks: A Recap

Generative AI stands as one of the most influential technologies in the modern business landscape. It offers unprecedented capabilities in content generation, decision-making processes, and customer engagement, reshaping industries and creating new opportunities for innovation and efficiency. In this comprehensive exploration, we recap the multifaceted benefits and potential drawbacks of integrating Generative AI into business operations, considering its impact on productivity, innovation, and the overarching economic landscape.

The Multitude of Business Benefits

Enhanced Creativity and Content Production

Generative AI technologies are redefining the boundaries of content creation. By utilizing advanced algorithms, businesses can generate written content, images, and even videos that cater to specific audiences with minimal human intervention. This capability significantly accelerates the content production process, enabling organizations to respond rapidly to market demands and maintain a consistent presence across multiple media platforms.

Personalization at Scale

In the realm of marketing and customer service, Generative AI allows for personalized experiences at an unprecedented scale. AI-driven systems can analyze customer data to deliver tailored recommendations, advertisements, and communications. This level of personalization, once a resource-intensive endeavor, can now be achieved efficiently, fostering deeper customer relationships and enhancing brand loyalty.

Innovation in Product and Service Development

Generative AI also serves as a catalyst for innovation. In product development, for example, AI algorithms can simulate and evaluate countless design variations, thereby aiding in the creation of novel products that meet specific customer needs or create entirely new market niches. Similarly, in service delivery, AI can generate innovative solutions to complex problems, thereby differentiating a company's offerings in a competitive marketplace.

Automation of Routine Tasks

Another significant benefit of Generative AI is the automation of routine and repetitive tasks. From drafting emails to generating reports, AI can free up human resources, allowing employees to focus on more strategic and creative tasks that add greater value to the business.

Data-Driven Decision-Making

Generative AI excels in processing vast amounts of data to inform decision-making. In sectors such as finance and logistics, AI models can predict market trends, optimize supply chains, and enhance risk management, leading to more informed and timely business decisions.

The Potential Drawbacks

Ethical and Privacy Concerns

The power of Generative AI to create and manipulate content raises ethical questions, particularly concerning authenticity and ownership. The generation of deepfakes or misleading content can have severe implications for individuals and society. Additionally, the reliance on extensive data to fuel AI systems brings about privacy concerns, as sensitive information must be handled with utmost care to avoid breaches and misuse.

Dependency and Reduced Human Creativity

An over-reliance on AI for content generation may lead to a reduction in human creativity. There is a potential risk that as AI becomes more adept at creating content, the value and development of human-driven creativity could diminish, potentially stifling innovation in the long term.

Job Displacement

Automation through AI could lead to displacement in certain job sectors. As AI takes over routine tasks, there may be a decrease in demand for human labor in these areas, which could have broader socioeconomic impacts, such as increased unemployment and the need for re-skilling of the workforce.

Implementation Costs and Complexity

While the long-term benefits of AI are clear, the initial implementation can be costly and complex. Businesses must invest in infrastructure, acquire the right talent to manage AI systems, and ensure that there is a strategic alignment with the company's goals, which can be a significant hurdle, particularly for small- and medium-sized enterprises.

Quality Control and Brand Risk

Generative AI systems are only as good as the data they are trained on, and they may produce content that is not aligned with a company's brand voice or standards. Ensuring quality and consistency requires ongoing oversight and fine-tuning of AI models, which can be a continuous resource investment.

Conclusion

Generative AI presents a compelling dichotomy for businesses. On the one hand, it promises efficiency, personalization, and innovation; on the other hand, it carries ethical, economic, and operational challenges that must be addressed. As organizations increasingly adopt Generative AI, they must navigate these benefits and drawbacks carefully, implementing robust frameworks to maximize the potential of AI while mitigating its risks. Ultimately, the goal is to harmonize the strengths of Generative AI with the creative and strategic capabilities of the human workforce, thus fostering a synergistic environment where technology and human expertise drive collective progress and sustainable business growth.

Strategies for Implementation

In the contemporary digital era, the implementation of Generative AI within a business setting is a strategic endeavor that requires meticulous planning, a nuanced understanding of the technology, and a forward-thinking approach to integration. As companies seek to harness the benefits of Generative AI—ranging from enhanced content creation to sophisticated data analysis—they must adopt comprehensive strategies to ensure successful deployment. Here we delve into the multifaceted strategies for the implementation of Generative AI, exploring the stages from foundational preparation to advanced integration, and ongoing optimization.

Foundational Preparation: Laying the Groundwork for Generative AI

Comprehending the Technology

Before embarking on the implementation of Generative AI, it is imperative for an organization to develop a thorough understanding of the technology. This entails not only grasping the basic principles of machine learning, neural networks, and data processing but also recognizing the specific nuances of Generative AI applications such as natural language processing, computer vision, and predictive analytics.

Strategic Alignment

The adoption of Generative AI must be closely aligned with the business's strategic objectives. Leaders should identify key areas where AI can drive value, be it through boosting operational efficiency, enhancing customer engagement, or fostering innovation. A clear strategic vision will guide the prioritization of AI initiatives and ensure that the technology serves to advance the company's long-term goals.

Stakeholder Engagement

Successful implementation requires buy-in across all levels of the organization. Engaging stakeholders—from executives to frontline employees—in understanding the benefits and changes associated with Generative AI is crucial. This could involve educational workshops, demonstrations, and discussions to address concerns and highlight the potential for AI to augment human capabilities.

Infrastructure and Talent Acquisition

Building or Acquiring the Necessary Infrastructure

Implementing Generative AI requires a robust technological infrastructure capable of handling large datasets and supporting complex AI algorithms. This might involve upgrading existing IT systems, migrating to cloud services, or investing in specialized hardware. The choice of infrastructure should consider scalability, security, and compliance with industry regulations.

Acquiring and Developing Talent

Having the right talent in place is essential for navigating the complexities of AI. This may include recruiting data scientists, AI specialists, and domain experts or investing in training programs to upskill existing staff. A multidisciplinary team can facilitate the effective development, management, and oversight of AI systems.

Data Management and Governance

Data Acquisition and Analysis

Generative AI is data-driven, making the acquisition of high-quality data a cornerstone of successful implementation. Organizations need to establish processes for collecting, cleaning, and structuring data to train AI models effectively. Additionally, they must ensure ethical data practices, protecting user privacy and adhering to data protection laws.

Governance Framework

Developing a governance framework is key to overseeing AI operations. This includes setting policies for data usage, model training, and output evaluation. Governance also encompasses the ethical considerations of AI deployment, ensuring that generated content and decisions reflect the company's values and societal norms.

Incremental Implementation and Iterative Development

Pilot Projects

Starting with small-scale pilot projects can allow businesses to test the waters with Generative AI. These projects serve as a testing ground to assess the technology's impact, identify potential issues, and gather insights that can inform broader implementation efforts.

Iterative Development

AI systems are not static; they require ongoing development and refinement. An iterative approach, incorporating feedback loops and continuous improvement, ensures that AI models remain accurate, relevant, and aligned with evolving business needs.

Integration and Scaling

Cross-Functional Integration

Generative AI should not operate in a silo but be integrated across various business functions. This ensures that AI-generated insights and content are leveraged across the organization, from marketing and sales to customer service and product development.

Scaling AI Initiatives

As the organization becomes more comfortable with Generative AI, scaling initiatives to capitalize on its full potential becomes possible. This may involve expanding the range of applications, increasing the volume of data processed, or extending AI capabilities to new business areas.

Monitoring, Evaluation, and Optimization

Performance Monitoring

Implementing monitoring systems to track the performance of AI applications is vital. Key performance indicators should be established to measure the effectiveness of AI in achieving desired outcomes.

Evaluation and Feedback Mechanisms

Regular evaluation and solicitation of feedback from users and stakeholders contribute to the calibration of AI systems. This feedback loop can help identify areas for improvement and ensure that the AI remains aligned with user needs and expectations.

Continuous Optimization

AI is an evolving field, and staying abreast of the latest developments is crucial for maintaining a competitive edge. Continuous optimization may involve updating AI models with new data, incorporating advanced algorithms, or adopting emerging technologies.

Summary

The strategic implementation of Generative AI is a complex but potentially transformative process for businesses willing to invest in its potential. It requires a comprehensive approach, encompassing a deep understanding of the technology, alignment with business strategy, robust infrastructure, skilled talent, and rigorous governance. By adopting an incremental, iterative approach to development and integration, organizations can successfully implement Generative AI into their businesses.

CHAPTER 15

The Evolving World of Generative AI

In the rapidly advancing domain of technology, Generative AI is emerging as a pivotal force, steering us into a new epoch of innovation and creativity. This branch of AI, renowned for its capability to create and innovate, is moving beyond traditional realms of data processing to actively generate new forms of content and solutions. This shift marks a significant transformation in the technological world, promising a future where AI's influence extends across diverse sectors, redefining how we interact with technology and perceive its potential. Here we explore the various next-generation technologies in Generative AI, which are set to revolutionize our interactions in both the digital and physical realms.

At the forefront of this transformation are the advancements in multimodal AI systems. These systems represent a convergence of different sensory inputs, including vision, language, and sound, within a single AI framework. This fusion enables AI to process and generate complex information, mirroring the human ability to engage with the world using multiple senses simultaneously. Future developments in multimodal AI are expected to introduce pioneering capabilities, transforming the way AI interprets and interacts with its environment.

These advancements in multimodal AI encompass more sophisticated mechanisms for fusing various sensory inputs, leading to richer, more nuanced contextual understanding. For instance, future AI systems could have the capability to process a film in its entirety—analyzing dialogue, soundtracks, and visual cues—and then synthesize a comprehensive understanding that captures not just the narrative but also the thematic subtleties.

Additionally, the dynamic generation of content across different modalities represents a significant leap forward. Generative AI could, for example, create three-dimensional models based on verbal descriptions or produce educational content

© Irena Cronin 2024
I. Cronin, *Understanding Generative AI Business Applications*, https://doi.org/10.1007/979-8-8688-0282-9_15

that adapts dynamically to a student's learning style, presenting information visually, textually, or aurally as needed. This versatility in content generation paves the way for AI to be more integrative and adaptive to human needs and preferences.

Another exciting avenue is the enhanced interaction of AI systems with the physical environment. Future multimodal AI systems are likely to become more adept at interpreting and engaging with their surroundings. This could be particularly transformative in robotics and AR, where AI could enable systems to navigate and interact with their environment with a level of precision and understanding that closely mirrors human capabilities. These systems could combine visual data, tactile feedback, and semantic understanding to perform complex tasks and offer intuitive user experiences.

Furthermore, the integration of AI with AR and VR technologies is poised to create more immersive and interactive experiences. Future AI systems could generate virtual environments that respond to the user's verbal cues, gaze, and gestures, adjusting the virtual scenario in real time to create a fully interactive and personalized experience. This integration promises to blur the lines between the digital and physical worlds, offering immersive experiences that were once the realm of science fiction.

The horizon of Generative AI is marked by a series of transformative advancements that promise to redefine our interaction with technology in profound ways. From multimodal AI systems and dynamic content generation to enhanced interactions with the physical environment and the integration with AR and VR technologies, these advancements are set to revolutionize a wide array of fields. As we embrace these next-generation technologies, it is crucial to navigate their development thoughtfully, ensuring that they are used to enhance human potential and well-being.

Next-Generation Technologies on the Horizon

As we venture deeper into the 21st century, the technological landscape is being reshaped by the rapid evolution of Generative AI. This branch of AI is not just about understanding or processing information but about creating it. The horizon is bright with a spectrum of next-generation technologies that promise to further revolutionize the field. Here we examine the anticipated advancements and innovations in Generative AI that are poised to unfold.

Advancements in Multimodal AI Systems

Multimodal AI systems represent the convergence of different sensory modalities such as vision, language, and sound within a single AI framework. This enables the AI to process and generate complex information that encompasses multiple types of data simultaneously, much like how humans engage with the world using multiple senses. As we look to the future, advancements in multimodal AI systems are anticipated to be a key area of growth within the field of Generative AI. These advancements are expected to bring forth several pioneering capabilities and applications.

Fusion of Sensory Inputs for Richer Contextual Understanding

Future multimodal systems will likely feature more sophisticated mechanisms for fusing inputs from various modalities, such as combining visual, auditory, and textual information to understand context at a deeper level. This could mean an AI system that can watch a movie; process the dialogue, soundtrack, and visual cues; and then create a comprehensive summary that captures not just the plot but the thematic nuances.

Dynamic Content Generation Across Modalities

Advancements in Generative AI will enable the dynamic creation of content across different modalities. For example, an AI could generate a 3D model from a verbal description and then produce an accompanying narrative or even simulate the acoustic environment that the model would have. In the same vein, AI could create educational content that dynamically adapts to a student's learning style, presenting information visually, textually, or aurally as needed.

Improved Interaction with Physical Environments

As multimodal AI systems evolve, they will become more adept at interpreting and interacting with the physical world. This could be particularly transformative in robotics, where AI could enable robots to navigate and manipulate their environment with a level of precision and understanding that closely mimics human interaction. Such systems could combine visual data, tactile feedback, and semantic understanding to perform complex tasks.

AR and VR Integration

Multimodal AI will play a crucial role in the advancement of AR and VR, leading to more immersive and interactive experiences. Future AI systems could generate virtual environments that respond to the user's verbal cues, gaze, and gestures, adjusting the virtual scenario in real time to create a fully interactive and personalized experience.

Cross-Modal Data Synthesis and Translation

Another exciting frontier is cross-modal data synthesis and translation. Imagine an AI that can translate a piece of music into a visual artwork that captures the mood and rhythm of the composition or an AI that can create a descriptive narrative from a silent film. These systems would not only translate across modalities but also enrich the content by adding layers of interpretation and meaning that are not present in the original.

Emotionally Responsive Systems

Advancements in affective computing will likely be integrated into multimodal AI systems, enabling them to respond to human emotions in a more nuanced way. This could see AI systems that adjust the tone and style of generated content based on the emotional cues detected from a user's voice, facial expressions, or body language, leading to more empathetic and engaging interactions.

The future of multimodal AI systems lies in their ability to seamlessly integrate and synthesize complex forms of data, leading to more intelligent, intuitive, and interactive applications. These systems will not only enhance our ability to generate creative and insightful content but will also revolutionize how we interact with technology, each other, and the world around us. As we move forward, it is imperative to foster advancements in this field responsibly, ensuring that these powerful capabilities are used to enhance human potential and well-being.

Enhanced Learning Efficiency

In the burgeoning era of AI, the paradigm is shifting rapidly from data-intensive to efficiency-focused models. Next-generation Generative AI is at the forefront of this shift, heralding a new age of enhanced learning efficiency. Here we explore how advancements in AI learning techniques are set to revolutionize the field by requiring significantly fewer data points to discern complex patterns and produce high-quality output. We examine the mechanisms of transfer learning, few-shot learning, and meta-learning and consider the broader implications of these advancements for democratizing AI and mitigating environmental impact.

Transfer Learning: Capitalizing on Preexisting Knowledge

Transfer learning stands as a beacon of efficiency in the AI learning process. It involves taking a model that has been trained on one task and repurposing it for another related task. This approach capitalizes on the knowledge that the model has already acquired, thereby reducing the need for extensive retraining from scratch. For Generative AI, this means that a model trained to generate text in one domain can quickly adapt to generate relevant content in another, vastly different domain, with minimal additional input. This not only accelerates the training process but also reduces the computational resources required, making AI more accessible.

Few-Shot Learning: Mastery with Minimal Examples

Few-shot learning is another promising avenue that enables AI models to understand and replicate complex patterns after being exposed to only a small number of examples. This technique is akin to a human artist who can grasp the essence of a style after seeing just a few paintings. In the context of Generative AI, few-shot learning will empower systems to create content that is coherent and contextually relevant with little initial data, thereby opening up possibilities for AI applications in data-scarce environments and for languages or dialects that have limited written resources.

Meta-Learning: The AI that Learns to Learn

Meta-learning, or learning to learn, represents a meta-cognitive leap in AI training. This involves designing AI models that can improve their learning algorithms over time, enabling them to become more proficient with each task they perform. Such models would not only learn more efficiently from new data but would also refine their learning processes based on past experiences. For Generative AI, meta-learning equates to an ongoing enhancement of creativity and relevance, as the AI becomes increasingly adept at pattern recognition and content generation across diverse domains.

Democratizing AI

These advanced learning techniques are poised to democratize AI by lowering the barriers to entry. With reduced requirements for data and computational power, AI technology becomes more accessible to smaller enterprises and individual developers. This democratization could unleash a wave of innovation as more people are able to develop and deploy AI solutions tailored to specific local or niche problems, driving progress in areas that have been underserved by technology.

Addressing Environmental Impact

The efficiency of next-generation AI learning has significant environmental implications. Traditional deep learning models require vast amounts of computational power, which in turn demands considerable energy, often derived from carbon-intensive sources. By reducing the computational load necessary for training AI models, these new techniques can help decrease the carbon footprint of AI research and deployment, aligning the growth of AI with sustainability goals.

The shift toward enhanced learning efficiency in Generative AI represents a transformative phase in the evolution of AI. As next-generation models become more adept at learning from limited data through transfer learning, few-shot learning, and meta-learning, we stand on the cusp of a more inclusive, innovative, and environmentally conscious AI landscape. These advances will enable us to harness the power of AI more effectively and responsibly, ensuring that its benefits are widely distributed, and its applications are sustainable for the planet. As we embrace these technologies, it is incumbent upon us to navigate their development with foresight and a commitment to the ethical deployment of AI.

Integration with Quantum Computing

The integration of quantum computing with Generative AI is a development poised to redefine the boundaries of computational science and AI. This symbiotic relationship holds the promise of harnessing the peculiar and powerful properties of quantum mechanics to amplify the capabilities of AI systems, especially those dedicated to generative tasks. The potential of this integration encompasses the optimization of neural networks, the generation of complex simulations, and the resolution of intricate problems that are currently beyond the reach of classical computing infrastructures.

Quantum Computing: The Catalyst for Computational Revolution

Quantum computing operates fundamentally differently from classical computing by utilizing quantum bits or qubits, which can exist in multiple states simultaneously through the phenomenon known as superposition. This, combined with entanglement, another quantum mechanic principle, allows quantum computers to perform many calculations at once, offering a dramatic increase in processing power. For Generative AI, this means the ability to analyze and learn from data at speeds and scales that were previously unimaginable.

Enhanced Neural Network Optimization

The application of quantum algorithms to neural networks could result in optimization techniques that drastically reduce the time required to train AI models while potentially increasing their accuracy. Quantum computing could enable Generative AI to swiftly adjust neural network parameters, even for extremely complex networks, thereby enhancing the model's performance on generative tasks such as creating lifelike images, composing music, or writing coherent and contextually relevant narratives.

Complex Simulations and Problem-Solving

One of the most anticipated applications of quantum-enhanced Generative AI is in the field of simulations. Quantum computing could enable the simulation of molecular and quantum systems with high fidelity, a task that is computationally intensive and impractical with current classical computers. In logistics, quantum algorithms could optimize routes and distribution networks more efficiently, handling a multitude of variables and constraints with ease.

Advancing Materials Science

In materials science, the implications are profound. The ability of quantum computers to model complex molecular structures could accelerate the discovery of new materials, leading to advancements in everything from batteries and electronics to pharmaceuticals. The integration of quantum computing with Generative AI could automate and innovate the design process, predicting properties of materials that have yet to be created.

Revolutionizing Logistics

Quantum computing could transform logistics, a field that depends on the optimization of numerous variables to improve efficiency and reduce costs. Generative AI, powered by quantum computing, could design systems that not only predict and react to logistical challenges but also evolve supply chain mechanisms to unprecedented levels of efficiency.

Ethical and Security Considerations

With great power comes great responsibility, and the quantum leap in AI capabilities necessitates a discussion on ethics and security. Quantum computing could, in theory, break many of the encryption methods currently in use, which raises significant data privacy and security concerns. Thus, the integration of quantum computing with AI must be accompanied by advancements in quantum cryptography to protect against these potential vulnerabilities.

The integration of quantum computing with Generative AI represents a leap forward into a new computational era. It offers the prospect of solving some of the most complex and challenging problems in science and industry. As we stand at the precipice of this technological revolution, it is essential to guide its development with a keen eye on the ethical and security implications, ensuring that the power of quantum computing is harnessed for the benefit of society and the betterment of human life. As these technologies converge, they will unlock new possibilities, catalyze discovery, and revolutionize countless fields, marking the dawn of a new chapter in the annals of human ingenuity.

AI-Driven Personalization

The trajectory of technological advancement has increasingly inclined toward personalization, with AI-driven personalization technologies at the helm, steering us toward a future where every digital interaction is tailored to the individual. Generative AI is set to play a pivotal role in this shift, offering bespoke content and experiences that adapt to the unique preferences and behaviors of each user. The scope of these technologies spans various domains, from education to entertainment, commerce, and healthcare, promising a transformative impact on both individual experiences and broader societal patterns.

Personalized Learning Curriculums

In the realm of education, AI-driven personalization heralds a revolution in how curricula are designed and delivered. Generative AI has the potential to create customized learning pathways that adapt to the pace, style, and interests of individual learners. This approach could dismantle the one-size-fits-all model of education, allowing for a more inclusive system that caters to diverse learning needs and optimizes educational outcomes. For instance, a student struggling with mathematics might receive additional problems that are calibrated to their level of understanding, along with explanatory content that aligns with their preferred learning modalities.

Customized Entertainment and Shopping Experiences

The entertainment industry is poised to be radically transformed by AI that can generate content based on individual tastes and past interactions. Streaming services could offer not just recommendations but entire films or music albums created on-demand to suit the mood and preferences of the viewer or listener. Similarly, the retail and e-commerce sectors will benefit from AI that can anticipate purchasing preferences, recommend products, and even customize the shopping interface to reflect the user's behavior patterns, significantly enhancing the consumer experience.

Healthcare Personalization

Perhaps one of the most impactful applications of AI-driven personalization lies in healthcare. Generative AI could develop treatment plans that are deeply personalized, taking into account an individual's genetic profile, medical history, and lifestyle factors. Such tailored healthcare could lead to better patient outcomes and a more efficient healthcare system. For example, AI could suggest a nutrition plan that not only considers an individual's health conditions but also their taste preferences and daily routines, thereby increasing the likelihood of adherence.

Extending to Lifestyle and Daily Interactions

The influence of AI-driven personalization is also set to permeate everyday life. Smart homes could adjust lighting, temperature, and even scent based on the presence and mood of the occupants. Personalized news feeds could become more nuanced, presenting a balance of topics that align with the user's interests while also introducing new subjects to broaden their horizons without overwhelming them.

Challenges and Considerations

While the benefits are clear, the rise of AI-driven personalization also presents challenges, particularly in privacy and ethical decision-making. As AI systems collect and analyze vast amounts of personal data to provide these tailored experiences, ensuring the security and consent of users becomes paramount. Additionally, there is a risk of creating "filter bubbles," where the AI's personalization algorithms might limit exposure to diverse perspectives and experiences, potentially reinforcing biases and narrowing personal growth.

The surge in AI-driven personalization technologies promises a future where each individual's digital and physical environments are uniquely their own, created and curated by the intelligent synthesis of their data. The vast potential of Generative AI to tailor content and experiences heralds a new era of personal engagement and satisfaction across various sectors. As we advance toward this highly personalized future, it is crucial to address the accompanying challenges with robust ethical frameworks and privacy safeguards, ensuring that AI-driven personalization enriches the human experience while respecting individual autonomy and diversity.

AI and AR Convergence

The Convergence of AI and AR: Reshaping Interaction with the World

The technological landscape is on the cusp of a transformative integration with the convergence of Generative AI and AR technologies. This fusion is set to redefine human interaction with the immediate environment by superimposing a digitally enriched layer onto the physical realm. The potential of this convergence extends across various fields, offering innovative ways to engage with information, education, and entertainment.

Redefining Educational Experiences

In the educational sector, the amalgamation of Generative AI with AR can revolutionize traditional learning methodologies. By leveraging the real-time, interactive capabilities of Generative AI, educational tools can morph into dynamic platforms that adapt to the learner's context. Imagine students exploring historical sites with AR that vividly reconstructs the past, all while AI tailors the information to each student's learning progress and interests. Such contextually enriched experiences can make learning more engaging and effective, bridging the gap between theoretical knowledge and practical understanding.

Transforming Entertainment

The entertainment industry stands to gain immensely from the AI and AR convergence. Video games, movies, and virtual tours can become truly immersive experiences, with AI-generated content that evolves in response to the user's interactions. AR could bring characters and environments to life in the user's living room, with storylines that adapt and change, offering a personalized narrative journey. This level of immersion could redefine the very nature of storytelling and audience engagement.

Enhancing User Interactivity

User interactivity with digital content will reach new heights as AR devices become more sophisticated and Generative AI more nuanced in its content creation. Through AR glasses or mobile devices, users could interact with virtual objects and information overlaid on the real world in a seamless and intuitive manner. Generative AI could instantly create these virtual objects based on the user's current task or environment, providing assistance or adding to the richness of the user's reality.

Augmenting Work Environments

In professional settings, the convergence can enhance productivity and collaboration. For architects and engineers, AR can project AI-generated models onto physical spaces, allowing for real-time visualization and modification. Surgeons could use AR to overlay medical imaging on patients during procedures, with AI assisting by providing predictive analytics and procedural guidance.

Challenges in the Path Forward

Despite its potential, the path to harmonizing AI with AR involves significant challenges. Issues of privacy, data security, and the need for robust infrastructure must be addressed. There is also the challenge of ensuring that the content generated by AI is accurate, relevant, and appropriate for all ages and backgrounds.

The convergence of AI and AR technologies promises a future where the boundaries between the digital and physical worlds become increasingly blurred. As Generative AI becomes more adept at creating responsive, context-aware content, and AR technologies advance in projecting this content onto our world, the possibilities for innovation seem boundless. This synergy has the potential to enrich our daily lives, transforming how we learn, work, and entertain ourselves. As we navigate toward this integrated future, it is essential to foster responsible development to fully realize the benefits while mitigating potential risks.

Ethical AI and Governance

The burgeoning power of Generative AI ushers in a need for conscientious oversight through ethical frameworks and governance models. As AI technologies grow more sophisticated, they intersect increasingly with societal norms, necessitating a careful balance between innovation and responsibility. The task at hand is not merely technical; it is profoundly moral and administrative, involving the creation of systems that ensure AI is employed for the common good while safeguarding individual rights.

Building Robust Ethical Frameworks

The establishment of ethical frameworks in AI involves setting principles that dictate the fair and equitable development and deployment of AI systems. These frameworks are vital for addressing inherent issues such as algorithmic bias, which can perpetuate societal inequalities if left unchecked. Ethical AI must strive to be inclusive, drawing from diverse datasets and considering a multiplicity of perspectives to mitigate biases.

In addition, ethical frameworks must define the parameters of privacy, particularly as AI systems become more adept at synthesizing personal data to generate insights and content. It is crucial to respect the autonomy of individuals, ensuring that data is not only collected and used with consent but also safeguarded against breaches that could compromise personal information.

Governance Models for AI

Governance models play a pivotal role in the practical implementation of ethical principles. These models involve policies and regulations that guide the development and application of AI technologies. Governance encompasses the enforcement of standards, the oversight of AI practices, and the establishment of accountability mechanisms for when things go wrong.

One aspect of governance is ensuring transparency in AI operations. The "black-box" nature of many AI systems, where the decision-making process is opaque, is a challenge to public trust. Opening up these systems for scrutiny, perhaps through explainable AI (XAI) practices, is essential for fostering confidence among users and stakeholders.

Preventing Misuse of AI

The potential for the misuse of AI, whether through the creation of deepfakes or the manipulation of public opinion, is a pressing concern. Ethical frameworks and governance models must anticipate such risks and establish preventative measures. This might involve the development of technologies that can detect AI-generated fabrications or the enactment of laws that penalize malicious uses of AI.

Global Cooperation on AI Ethics and Governance

AI technology knows no borders, and thus, ethical frameworks and governance models must be developed with a global perspective. International cooperation is crucial in setting cross-border standards and practices. A global dialogue can help ensure that AI benefits all of humanity and that the risks associated with AI do not disproportionately affect certain regions or demographics.

The evolution of Generative AI into more advanced and powerful forms brings with it a profound responsibility to guide its development with ethical frameworks and governance models. These measures are imperative for ensuring that AI technologies are used in a manner that is responsible, equitable, and transparent. They provide the foundation for maintaining public trust and for harnessing the vast potential of AI in a way that respects individual rights and promotes the welfare of society at large. As we continue to innovate in the field of AI, the principles of ethical AI and governance must remain at the forefront, steering the course of technological advancement toward a future that is just and beneficial for all.

Regulatory Landscape

The rapid advancement of Generative AI technologies presents a complex challenge for regulators around the globe. The task of governing such dynamic and evolving systems requires a forward-looking regulatory landscape that is both adaptive and proactive. As Generative AI continues to permeate various sectors, from creative industries to healthcare and finance, the need for a robust regulatory framework becomes increasingly critical to address the ethical, legal, and socioeconomic implications of this technology. An example of a body that is involved in guidelines is the World Economic Forum (WEF). The AI Governance Alliance of the WEF serves as a platform where leaders from industry, government bodies, academic circles, and civil organizations come together. It recently disclosed responsible and ethical guidelines for Generative AI. As a note, given that Generative AI technology is making such fast progress, it is important to keep track of the most updated policies in the area.

Adaptive Legal Frameworks

The future regulatory landscape will likely consist of adaptive legal frameworks designed to evolve alongside Generative AI technologies. These frameworks must be flexible enough to accommodate new developments while providing clear guidelines on permissible uses of AI. Legislators may adopt a principle-based approach to regulation, which sets out broad objectives such as fairness, accountability, and transparency, rather than prescriptive rules that could quickly become outdated.

Standardization of Practices

Regulation will play a key role in standardizing practices within the field of Generative AI. This includes the standardization of datasets to prevent biases in AI-generated content and the development of benchmarks for evaluating the quality and safety of AI outputs. Standardization ensures a level playing field and can facilitate the interoperability of AI systems across different platforms and industries.

Ethics Committees and Oversight Bodies

It is probable that ethics committees and oversight bodies will be established to monitor the deployment of Generative AI. These bodies could be tasked with the continuous review of AI applications to ensure compliance with ethical standards. By conducting audits and assessments, they would provide oversight on how AI systems are designed, what data they are using, and how their outputs are influencing the public domain.

Privacy Regulations

Privacy regulations will form a cornerstone of the Generative AI regulatory landscape. As Generative AI systems often rely on large amounts of personal data to function, ensuring the privacy and security of this data is paramount. Regulations like the General Data Protection Regulation (GDPR) in the European Union may serve as a template, emphasizing the importance of consent, data minimization, and the right to be forgotten.

Intellectual Property Rights

The unique ability of Generative AI to create original content will prompt a reexamination of intellectual property rights. The regulatory landscape will need to address questions about the ownership of AI-generated works, the rights of those whose data was used to train the AI, and how to attribute contributions between human creators and AI systems.

Combating Misuse

Preventing the misuse of Generative AI will be an essential focus of future regulations. This includes establishing legal recourse against the creation and distribution of deepfakes and other deceptive materials. Lawmakers will need to balance the protection of individuals and society with the freedom of expression and innovation.

Global Coordination

Given the borderless nature of technology and data, international coordination in regulatory approaches will be crucial. Global treaties and agreements may be necessary to manage the cross-border challenges posed by Generative AI, ensuring consistent standards for responsible AI usage worldwide.

The regulatory landscape for Generative AI is set to be a complex and ever-evolving terrain. It will necessitate a blend of adaptive legal frameworks, standardization, privacy protection, intellectual property rights clarification, and mechanisms to combat misuse. As we navigate the future of Generative AI, a collaborative approach involving stakeholders from governments, industry, academia, and civil society will be instrumental in crafting regulations that not only protect against risks but also foster an environment where innovation and public interest thrive in concert. The task ahead is formidable, but with thoughtful regulation, the promise of Generative AI can be realized in a manner that is beneficial and equitable for all.

Future Research Directions

As Generative AI continues its rapid evolution, it is essential to identify and explore future research directions that can further advance this field. The potential of Generative AI extends far beyond its current capabilities, offering vast prospects for innovation, creativity, and problem-solving. Here we explore the key areas where future research efforts are likely to be concentrated, highlighting the potential breakthroughs and challenges that lie ahead.

Enhancing Creativity and Diversity of AI Outputs

One primary area of research will focus on enhancing the creativity and diversity of outputs generated by AI systems. Current models are often limited by the data they are trained on, which can lead to repetitive or biased outputs. Future research will aim to develop AI systems capable of more original thought and diverse creative expressions. This could involve exploring new training methods or algorithmic structures that encourage more innovative and varied content generation.

Improving Multimodal Capabilities

As technology advances, the integration of multimodal capabilities in Generative AI will be a significant research area. This involves AI systems that can understand and generate content across multiple formats—such as text, image, audio, and video—in a cohesive manner. Enhancing these capabilities will allow for more sophisticated interactions between AI and users and open up new possibilities in fields like digital media, education, and entertainment.

Advancing Personalization Techniques

Another key research direction is the advancement of personalization techniques in Generative AI. Future research will likely explore ways to tailor AI-generated content more precisely to individual preferences, behaviors, and contexts. This personalization extends beyond consumer applications to include personalized learning systems, healthcare diagnostics, and therapies.

Ethical AI Development

The ethical development of Generative AI is an increasingly important area of research. As AI systems become more advanced, they raise complex ethical questions regarding bias, privacy, and autonomy. Future research must address these challenges, developing AI systems that are not only technically proficient but also ethically sound and socially responsible.

Interdisciplinary Collaboration

Generative AI research is also moving toward more interdisciplinary collaboration. Combining insights from fields such as psychology, neuroscience, and sociology with AI research can lead to a deeper understanding of human creativity and intelligence, which in turn can inspire new approaches in AI development.

Quantum AI Integration

The integration of quantum computing with AI presents an exciting frontier for research. Quantum computing offers the potential to process information at unprecedented speeds, which could revolutionize the capabilities of Generative AI. Exploring how these two technologies can work together will be a crucial area of study in the years to come.

AI in Climate Change and Sustainability

An increasingly important research direction is the application of Generative AI in addressing climate change and promoting sustainability. AI can be used to model complex environmental systems, optimize energy consumption, and develop new materials for clean energy technologies.

AI in Environmental Monitoring and Conservation

Generative AI can be pivotal in environmental monitoring and conservation efforts. Research could focus on developing AI systems that can predict environmental changes, assist in biodiversity conservation, and optimize resource utilization, contributing significantly to sustainability efforts.

AI for Social Good

Research in Generative AI can also be directed toward social good applications, such as developing solutions for global challenges like poverty, education, and disaster response. AI can play a key role in analyzing large datasets to identify patterns and solutions that can aid in policymaking and humanitarian efforts.

Augmented Reality and Virtual Reality Integration

Integrating Generative AI with AR and VR is an area ripe for exploration. This integration can significantly enhance user experiences by generating real-time, context-aware content within AR and VR environments. For example, AI could create immersive educational environments or complex virtual worlds for gaming, tailored to the user's interactions and preferences.

Generative AI in Healthcare Diagnostics and Treatment

Another critical area is the application of Generative AI in healthcare, particularly in diagnostics and personalized treatment planning. Research here could focus on developing AI models that can analyze medical images with greater accuracy or predict patient outcomes based on vast datasets, including genetic information. This approach could revolutionize personalized medicine, allowing for treatments tailored to the individual's genetic makeup and lifestyle factors.

Cognitive and Emotional Intelligence in AI

Enhancing cognitive and emotional intelligence in Generative AI systems is a fascinating research direction. This involves developing AI that can understand and respond to human emotions, making interactions more natural and empathetic.

Such advancements could transform AI from a tool into a more collaborative partner in various domains, including education, therapy, and customer service.

AI for Creative Industries

In the creative industries, research can explore how Generative AI can collaborate with artists, musicians, and designers, offering novel ways to create art, music, and designs. This collaboration can lead to new forms of creative expression and potentially uncover new artistic styles and genres.

Ethical AI Deployment in Diverse Cultural Contexts

Understanding and addressing the ethical implications of AI deployment in diverse cultural contexts is a crucial research area. This includes studying the impact of AI on different societal groups and ensuring that AI applications respect cultural sensitivities and norms.

Summary

In conclusion, the horizon of Generative AI is illuminated by a series of groundbreaking advancements that promise to reshape our technological landscape significantly. The evolution of multimodal AI systems and their ability to synthesize complex data across various sensory modalities are poised to enhance our interaction with digital and physical environments alike. These systems, with their enhanced learning efficiency, are set to democratize AI, making it accessible to a broader range of users while also addressing environmental concerns.

The integration of quantum computing with Generative AI hints at a future where computational boundaries are extended, allowing for the solving of complex problems and the generation of intricate simulations with unprecedented efficiency. Simultaneously, the rise of AI-driven personalization and its convergence with AR and VR technologies are set to revolutionize our educational, entertainment, and professional experiences, offering highly tailored and immersive interactions.

Moreover, the ethical AI and governance framework development alongside a dynamic regulatory landscape will play a crucial role in guiding these technologies responsibly. Ensuring these advancements align with societal norms and ethical standards is imperative for maintaining public trust and maximizing the potential benefits of Generative AI.

As we look to the future, these developments in Generative AI not only promise a transformation in various industries but also offer a glimpse into a world where technology works in harmony with human creativity and intelligence. It is a future that beckons with possibilities of enhanced human potential and well-being, rooted in technological innovation and responsible advancement.

Index

Printed in the United States
by Baker & Taylor Publisher Services

Printed in the United States
by Baker & Taylor Publisher Services